Mastering Essential Math Skills

Problem Solving

Richard W. Fisher

IT IS ILLEGAL TO PHOTOCOPY THIS BOOK

Problem Solving - Item #408

ISBN 10: 0-9666211-8-2 • ISBN 13: 978-0-9666211-8-1

Notes to the Teacher or Parent

What sets this book apart from other books is its approach. It is not just a math book, but a system of teaching math. Each daily lesson contains three key parts: **Review Exercises**, **Helpful Hints**, and **Problem Solving**. Teachers have flexibility in introducing new topics, but the book provides them with the necessary structure and guidance. The teacher can rest assured that essential math skills in this book are being systematically learned.

This easy-to-follow program requires only fifteen or twenty minutes of instruction per day. Each lesson is concise and self-contained. The daily exercises help students to not only master math skills, but also maintain and reinforce those skills through consistent review - something that is missing in most math programs. Skills learned in this book apply to all areas of the curriculum, and consistent review is built into each daily lesson. Teachers and parents will also be pleased to note that the lessons are quite easy to correct.

This book is based on a system of teaching that was developed by a math instructor over a thirty-year period. This system has produced dramatic results for students. The program quickly motivates students and creates confidence and excitement that leads naturally to success.

Please read the following "How to Use This Book" section and let this program help you to produce dramatic results with your math students.

How to Use This Book

This book is best used on a daily basis. The first lesson should be carefully gone over with students to introduce them to the program and familiarize them with the format. It is hoped that the program will help your students to develop an enthusiasm and passion for math that will stay with them throughout their education.

As you go through these lessons every day, you will soon begin to see growth in the student's confidence, enthusiasm, and skill level. The students will maintain their mastery through the daily review.

Step 1

The students are to complete the review exercises, showing all their work. After completing the problems, it is important for the teacher or parent to go over this section with the students to ensure understanding.

Step 2

Next comes the new material. Use the "Helpful Hints" section to help introduce the new material. Be sure to point out that it is often helpful to come back to this section as the students work independently. This section often has examples that are very helpful to the students.

Step 3

It is highly important for the teacher to work through the two sample problems with the students before they begin to work independently. Working these problems together will ensure that the students understand the topic, and prevent a lot of unnecessary frustration. The two sample problems will get the students off to a good start and will instill confidence as the students begin to work independently.

Step 4

Solutions are located in the back of the book. Teachers may correct the exercises if they wish, or have the students correct the work themselves.

Table of Contents

Review Exercises

Note to students and teachers: This section will include necessary review problems from all topics covered in this book. Here are some simple problems with which to get started.

1.   336
      27
   + 242

2.   752
    − 68

3.   725 + 242 + 163 =

4.   324
    x 6

5.   5,003
    x 6

6.   7 x 6,382 =

Helpful Hints	Bar graphs are used to compare information.	1. Read the title. 2. Understand the meaning of the numbers. Estimate, if necessary. 3. Study the data. 4. Answer the questions, showing work if necessary.

Use the information in the graph to answer the questions.

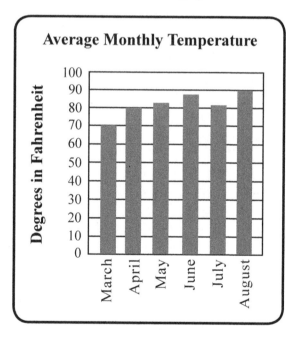

Average Monthly Temperature

S1. Which month had the second-lowest average temperature?

S2. How many degrees cooler was the average temperature in April than in August?

1. In which month was the average temperature 81°?

2. In which month did the average temperature drop from the previous month?

3. Which month had the second-highest average temperature?

4. How much warmer was the average temperature in August than in May?

5. For what month did the average temperature rise the most from the previous month?

6. Which months had average temperatures less than July's average temperature?

7. Which two month's average temperatures were the closest?

8. The coolest day in August was 77°. How much less than the average temperature was this?

9. What was the increase in average temperature from May to June?

10. Which months had an average temperature less than May?

1.

2.

3.

4.

5.

6.

7.

8.

9.

10.

Score

Review Exercises

1. 555
    77
    + 888

2. 703
    − 276

3. 642
    x   7

4. 42
    x 25

5. 47
    x 20

6. 76
    x 33

Helpful Hints

1. Read the title.
2. Understand the meaning of the numbers. Estimate, if necessary.
3. Study the data.
4. Answer the questions, showing work if necessary.

Use the information in the graph to answer the questions.

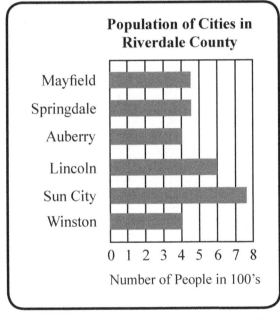

Population of Cities in Riverdale County

Mayfield
Springdale
Auberry
Lincoln
Sun City
Winston

0 1 2 3 4 5 6 7 8

Number of People in 100's

S1. Which two cities had the same population?

S2. What is the combined population of Mayfield and Lincoln?

1. Within 3 years, the population of Lincoln is expected to double.
What will its population be in 3 years?

2. How many more people live in Springdale than in Auberry?

3. What is the difference in population between the largest and smallest cities?

4. What is the total population of Riverdale County?

5. How many more people live in Sun City than in Mayfield?

6. To reach a population of 900, by how much must Mayfield grow?

7. What is the total population of the two largest cities?

8. How many people must move to Winston before its population is equal to that of Springdale?

9. Which city has nearly double the population of Auberry?

10. What is the total population of all cities whose population is less than 500?

1.

2.

3.

4.

5.

6.

7.

8.

9.

10.

Score

Review Exercises

1. 3,126
    x   5

2.   64
    x 23

3.   203
      x 47

4.   164
    x 23

5. 3 ⟌ 617

6. 5 ⟌ 236

Helpful Hints	Line graphs are used to show changes and relationships between quantities.	1. Read the title. 2. Understand the meaning of the numbers. Estimate if necessary. 3. Study the data. 4. Answer the questions, showing work if necessary.

Use the information in the graph to answer the questions.

S1. What was John's score on Test 5?

S2. How much better was John's score on Test 7 than on Test 3?

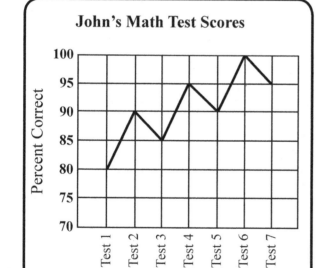

John's Math Test Scores

1. On which three tests did John score the highest?

2. What is the difference between his highest and lowest score?

3. Find John's average score by adding all his scores and dividing by the number of scores.

4. How many scores were below John's average score?

5. What are his two lowest scores?

6. What is the average of his highest and lowest scores?

7. How much higher was Test 4 than Test 5?

8. What was the difference between his highest score and his second lowest score?

9. How many test scores were improvements over the previous test?

10. Did John's progress generally improve or get worse?

1.
2.
3.
4.
5.
6.
7.
8.
9.
10.
Score

6

Review Exercises

1. 3,763
    472
    + 5,637

2. 5,016
    − 738

3. 435
    x 26

4. 7 ⟌ 1,407

5. 6 ⟌ 2,397

6. 8 ⟌ 1,445

Helpful Hints	1. Read the title. 2. Understand the meaning of the numbers. Estimate, if necessary. 3. Study the data. 4. Answer the questions, showing work if necessary.

Use the information in the graph to answer the questions.

Johnson's Music Shop Sales

Compact Discs   Cassettes

S1. In what months were the most cassettes sold?

S2. How many more compact discs than cassettes were sold in August?

1. What was the total number of compact discs sold in May and August?

2. How many more compact discs were sold in July than in August?

3. In which month were more cassettes sold than compact discs?

4. During which month did compact discs outsell cassettes by the most?

5. In September, how many more compact discs were sold than cassettes?

6. Which two months had the highest total sales?

7. What was the total number of compact discs and cassettes sold in September?

8. What was the increase in sales of compact discs between May and June?

9. What was the decrease in sales of compact discs between July and August?

10. What were the two lowest total sales months?

1.
2.
3.
4.
5.
6.
7.
8.
9.
10.
Score

Review Exercises

1. 664
    x 23

2. 5,000
    − 856

3. 435 + 75 + 61 + 42 =

4. 30 ⟌ 96

5. 50 ⟌ 765

6. 9 ⟌ 1,809

Helpful Hints	A circle graph shows the relationship between the parts to the whole and to each other.	1. Read the title. 2. Understand the meaning of the numbers. Estimate if necessary. 3. Study the data. 4. Answer the questions, showing work if necessary.

Use the information in the graph to answer the questions.

S1. What percent of the family budget is spent for food?

S2. After the car payment and house payment are paid, what percent of the budget is left?

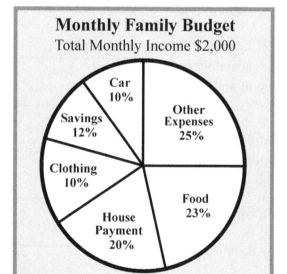

Monthly Family Budget
Total Monthly Income $2,000

Car 10%
Other Expenses 25%
Savings 12%
Clothing 10%
Food 23%
House Payment 20%

1. What percent of the budget is used to pay for food and clothing?

2. How much is spent on food each month? (Hint: Find 23% of $2,000.)

3. How many dollars are spent on clothing?

4. How many dollars was the house payment?

5. What percent of the budget did the three largest items represent?

6. What percent of the budget is for savings, clothing, and the car payment?

7. In twelve months, what is the total income?

8. What percent of the budget is left after the house payment has been made?

9. Which two items require the same part of the budget?

10. Which part of the budget would pay for medical expenses?

1.

2.

3.

4.

5.

6.

7.

8.

9.

10.

Score

8

Review Exercises

1. $2\overline{)76}$

2. $725 \div 5 =$

3. $80\overline{)905}$

4. $70\overline{)697}$

5. $70\overline{)8,196}$

5. $30\overline{)6,152}$

Helpful Hints	Circle graphs can be used to show fractional parts.	1. Read the title. 2. Understand the meaning of the numbers. 3. Study the data. 4. Answer the questions.

Use the information in the graph to answer the questions.

S1. What fraction of the day does Jane play?

S2. What fraction of Jane's day is used for school?

Jane's School Day

Sleep 9 hrs.
Home Work 2 hrs.
Chores 1 hr.
Play 4 hrs.
School 6 hrs.
Meals 2 hrs.

1. How many more hours does Jane sleep per day than play?

2. How many hours of homework does Jane have in a week (Monday through Friday)?

3. How many hours per day are school-related activities?

4. What fraction of the day is spent for school, homework, and chores?

5. What fraction of the day does Jane spend for school, sleep, and chores?

6. How many hours does Jane spend in school in 3 weeks?

7. If Jane goes to bed at 9:00 p.m., what time does she get up in the morning?

8. If school starts at 8:30 a.m., what time is school dismissed?

9. How many hours per week does Jane spend in school and on homework?

10. What fractional part of the day does Jane play and have meals?

1.	
2.	
3.	
4.	
5.	
6.	
7.	
8.	
9.	
10.	
Score	

Review Exercises

1.   364
       27
      256
    + 427

2.   7,000
    − 1,356

3.   365
    x 246

4.   22 ⟌463

5.   31 ⟌651

6.   18 ⟌379

| **Helpful Hints** | Picture graphs are another way to compare statistics. | 1. Read the title.
2. Understand the meaning of the numbers. Estimate, if necessary.
3. Study the data.
4. Answer the questions. |

Use the information in the graph to answer the questions.

Bikes Made By Street Bike Company

1986 🚲 🚲 🚲
1987 🚲 🚲 🚲 🚲
1988 🚲 🚲 🚲 🚲
1989 🚲 🚲 🚲 🚲 🚲 🚲
1990 🚲 🚲 🚲 🚲 🚲 🚲
1991 🚲 🚲 🚲 🚲 🚲 🚲 🚲 🚲

Each 🚲 represents 1,000 bikes.

S1. How many bikes were made in 1989?

S2. How many more bikes were made in 1991 than in 1988?

1. Which year produced twice as many bikes as 1986?

2. What was the total number of bikes that the company produced in 1990 and 1991?

3. Which two years did the company make the most bikes?

4. 1992 is reported to be double the production of 1988. How many bikes are to be produced in 1992?

5. What is the total number of biked produced in 1986 and 1991?

6. It cost $50 to make a bike in 1986. How much did the company spend making bikes that year?

7. The cost to make bikes jumped to $75 in 1991. How much did the company spend making bikes in 1991?

8. What is the difference in bikes produced in 1986 and 1991?

9. What is the total number of bikes made during the company's 3 most productive years?

10. One half of the bikes made in 1989 were ladies' style. How many ladies' bikes were made in 1989?

1.
2.
3.
4.
5.
6.
7.
8.
9.
10.
Score

10

Review Exercises

1. 3⟌52

2. 7⟌1,693

3. 30⟌697

4. 60⟌4,287

5. 22⟌7,695

6. 38⟌8,561

Helpful Hints	1. Read the title.

1. Read the title.
2. Understand the meaning of the numbers. Estimate if necessary.
3. Study the data.
4. Answer the questions, showing work if necessary.

Use the information in the graph to answer the questions.

S1. Which work week was the longest?

S2. How many hours shorter was the work week in 1960 than in 1990?

America's Work Week

1950
1960
1970
1980
1990

Each symbol represents 10 hours

1. How many hours long was the work week in 1970?

2. How many hours did the work week increase between 1980 and 1990?

3. If the average employee works 50 weeks per year how many hours did he work in 1950?

4. Which year's work week was approximately 38 hours?

5. Which years had the 2 shortest work weeks?

6. How many hours less was the work week in 1950 than 1980?

7. If the work week is 5 days, what was the average number of hours worked per day in 1950?

8. In 1990, if any employee decided to work 4 days per week, what would his average number of hours be per day?

9. Any work time over 40 hours is overtime. What was the average worker's weekly overtime in 1970?

10. What is the difference between the longest and the shortest work week?

1.
2.
3.
4.
5.
6.
7.
8.
9.
10.
Score

11

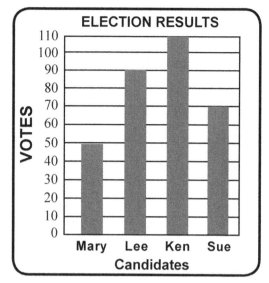

ELECTION RESULTS

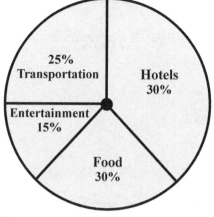

1. Which candidate got the most votes?

2. How many votes did Sue receive?

3. How many more votes did Lee get than Mary?

4. Together, how many votes did Lee and Ken receive?

5. Who got the third most votes?

6. What percent is spent on hotels?

7. What percent of the budget is spent on transportation?

8. What percent is spent altogether on transportation and hotels?

9. What is the second largest part of the budget?

10. What percent of the budget is used for entertainment?

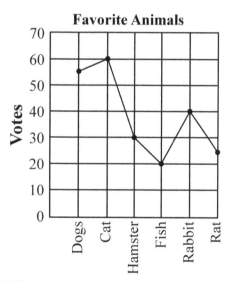

11. How many picked cats as their favorite animal?

12. What was the favorite animal?

13. What was the least favorite animal?

14. How many more people picked cats than hamsters?

15. What was the second favorite animal?

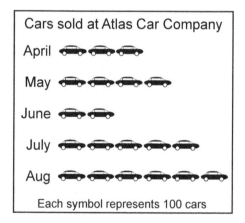

16. How many cars were sold in May?

17. Which month had the highest car sales?

18. How many more cars were sold in August than in June?

19. What was the total number of cars sold in April and June?

20. Which two months had the highest sales?

1.

2.

3.

4.

5.

6.

7.

8.

9.

10.

11.

12.

13.

14.

15.

16.

17.

18.

19.

20.

Score

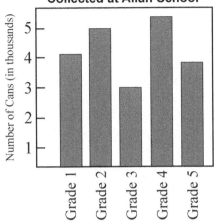

**Number of Aluminum Cans
Collected at Allan School**

1. Which grade collected the most cans?

2. How many more cans did grade 4 collect than grade 1?

3. How many cans did grades 4 and 5 collect altogether?

4. How many cans were collected in all?

5. Which grade collected the second most number of cans?

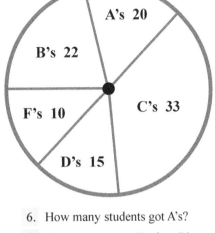
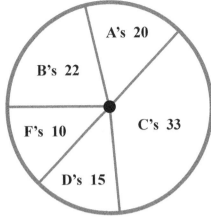

6. How many students got A's?

7. How many more C's than B's were there?

8. How many science students were there in all?

9. How many more C's than B's were there than A's and B's?

10. Which was the second largest group of grades?

Total Monthly Rainfall

11. How many inches of rain fell in February?

12. How many more inches of rain fell in January than in November?

13. Which month's rainfall increased the most from the previous month?

14. What was the total amount of rain for February, March, and April?

15. Which month's rainfall decreased the most from the previous month?

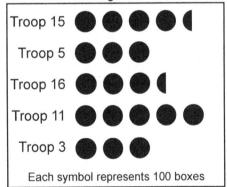

**Girl Scout Cookies
Sold During Fund-raiser**

Each symbol represents 100 boxes

16. How many boxes of cookies did Troop 11 sell?

17. Which troop sold the second most cookies?

18. How many more boxes did Troop 11 sell than Troop 5?

19. If Troop 3 sells three times as many boxes of cookies next year than they did this year, how much will they sell next year?

20. How many boxes of cookies did Troop 5 and Troop 15 sell altogether?

1.

2.

3.

4.

5.

6.

7.

8.

9.

10.

11.

12.

13.

14.

15.

16.

17.

18.

19.

20.

Score

13

Addition:
1. Line up on the right.
2. Place commas when necessary.
3. Add the ones first.
4. Regroup when necessary.
5. "Sum" means to add.

Subtraction:
1. Line up on the right.
2. Place commas when necessary.
3. Subtract the ones first.
4. Regroup when necessary.
5. "Difference" and "how much more" means to subtract.

Example:

$$
\begin{array}{r}
2\ \ 2\ 1 \\
7,654 \\
2,765 \\
793 \\
+\ \ \ \ 64 \\
\hline
11,276
\end{array}
$$

Example:

$$
\begin{array}{r}
6\ \ 11\ 9 \\
7,2\cancel{0}\cancel{1}3 \\
-\ 3,654 \\
\hline
3,549
\end{array}
$$

S1.
$$
\begin{array}{r}
743 \\
7,614 \\
16,321 \\
+\ 5,032 \\
\hline
\end{array}
$$

S2.
$$
\begin{array}{r}
5,001 \\
-\ 1,346 \\
\hline
\end{array}
$$

1.
$$
\begin{array}{r}
346 \\
25 \\
+\ 176 \\
\hline
\end{array}
$$

2.
$$
\begin{array}{r}
716 \\
-\ 143 \\
\hline
\end{array}
$$

3.
$$
\begin{array}{r}
4,216 \\
764 \\
+\ 5,123 \\
\hline
\end{array}
$$

4.
$$
\begin{array}{r}
3,732,246 \\
+\ 3,510,762 \\
\hline
\end{array}
$$

5.
$$
\begin{array}{r}
7,101 \\
-\ 1,436 \\
\hline
\end{array}
$$

6. Find the difference between 1,964 and 768.

7. $17,023 - 13,605 =$

8. Find the sum of 236, 742, and 867.

9. How much more is 763 than 147?

10. $72,163 + 16,432 + 1,963 =$

1.	
2.	
3.	
4.	
5.	
6.	
7.	
8.	
9.	
10.	
Score	

Multiplication:

1. Line up on the right.
2. Multiply ones first.
3. Multiply tens next.
4. Multiply hundreds last.
5. Add the products.
6. Place commas when necessary.

Example:

$$\begin{array}{r} 243 \\ \times\ 346 \\ \hline 1,458 \\ 9,720 \\ 72900 \\ \hline 84,078 \end{array}$$

Division:

1. Divide.
2. Multiply.
3. Subtract.
4. Divide again.
5. Remainders must be less than the divisor.

Example:

$$\begin{array}{r} 216\ \text{r}\,20 \\ 32\,\overline{)6,932} \\ -64\downarrow\downarrow \\ \hline 53 \\ -32 \\ \hline 212 \\ -192 \\ \hline 20 \end{array}$$

S1.
$$\begin{array}{r} 675 \\ \times\ 43 \end{array}$$

S2. $32\,\overline{)7,049}$

1.
$$\begin{array}{r} 76 \\ \times\ 3 \end{array}$$

2.
$$\begin{array}{r} 7,653 \\ \times\ 4 \end{array}$$

3.
$$\begin{array}{r} 627 \\ \times\ 36 \end{array}$$

4.
$$\begin{array}{r} 673 \\ \times\ 346 \end{array}$$

5. $3\,\overline{)425}$

6. $6\,\overline{)1697}$

7. $30\,\overline{)769}$

8. $40\,\overline{)3762}$

9. $42\,\overline{)8992}$

10. $28\,\overline{)1577}$

1.
2.
3.
4.
5.
6.
7.
8.
9.
10.
Score

Review Exercises

1.　　246
　　　　78
　　+ 912

2.　　7,562
　　−　399

3.　　709
　　x　3

4.　　　17
　　x　23

5.　2 ⟌ 117

6.　9 ⟌ 1,763

| **Helpful Hints** | 1. Read the problem carefully.
2. Find the important facts and numbers.
3. Decide what operation to use.
4. Solve the problem and label it with a word or short phrase. |

1.
2.
3.
4.
5.
Score

S1.　There are four classes with enrollments of 32, 35, 33, and 38.
　　　What is the total enrollment?

S2.　A family taking a vacation traveled 316 miles per day for five days.
　　　How many miles did they travel altogether?

1.　The attendance at the concert last year was 7,652. This year 8,043
　　attended. What was the increase in attendance?

2.　Six friends earned a total of 834 dollars. If they want to divide
　　the money equally, how much will each person receive?

3.　Mt. Everest is 29,028 ft. high. Mt. Whiting is 14,496 ft. high
　　How much taller is Mt. Everest than Mt. Whitney?

4.　A car traveled 330 miles in six hours. What was the average
　　speed per hour?

5.　A car's gas tank hold 22 gallons of gas. If the car can travel 31 miles
　　per gallon, how far can the car travel on a full tank of gas?

Review Exercises

1.   416
        x 22

2.   608
        x 342

3. $500 - 76 =$

4. $30\overline{)405}$

5. $60\overline{)7,123}$

6. $70\overline{)9,963}$

Helpful Hints

Look for key words:

1. "Total," "sum," "altogether," and "in all" usually mean addition or multiplication.

2. "How much more than," "difference," "what was the increase," and "how much was left" usually mean subtraction.

3. "Divide" and "average" usually mean division.

	1.
S1. A rope is 346 feet long. If it is divided into pieces two feet long, how many pieces will there be?	2.
S2. There are 195 students going on a field trip. If buses can hold 45 passengers, how many buses will be needed?	3.
	4.
1. Four friends went out to dinner and the bill came to $108. If they wanted to split the bill evenly, how much will each pay?	5.
2. A tank holds 55,000 gallons. If 33,500 gallons were removed, how many gallons would be left in the tank?	Score

3. Sixteen bales of hay weigh 2,000 pounds. What is the weight of each bale?

4. A factory can produce 55 cars per day. How many cars can it produce in 30 days?

5. A plane traveled 3,150 miles in seven hours. What was the average speed per hour?

Review Exercises

1. $30\overline{)44}$ 2. $30\overline{)532}$ 3. $30\overline{)7{,}132}$

4. $22\overline{)45}$ 5. $22\overline{)709}$ 6. $22\overline{)6{,}931}$

Helpful Hints

Remember:
1. Read the problem carefully.
2. Find the important facts and numbers.
3. Decide what operation to use.
4. Solve the problem and label it with a word or short phrase.

*Always label your answer with a word or short phrase.

1.	
2.	
3.	
4.	
5.	
Score	

S1. A man bought a used car for $2,045. After repainting it and making several repairs, he resold it for $2,750. What was his profit?

S2. Mr. Simmons bought a new dining room set for 936 dollars. If he pays for the set in 3 equal installments, how much is each payment?

1. In an election John received 2,693 votes. Kelly received 3,109 votes. How many more votes did Kelly receive than John?

2. A field in the shape of a triangle has sides 260 feet, 317 feet, and 289 feet. How many feet is it around the field?

3. A school is buying 45 books for $1,215. What is the price per book?

4. A farmer has 2,364 eggs. If he packs them in containers that hold 12 eggs, how many containers does he need?

5. A school has thirty classes. Each class contains 28 students. What is the total number of students in the school?

Review Exercises

1.
```
   236
    17
+  379
```

2.
```
  7,132
- 1,427
```

3. $73 + 96 + 74 + 62 =$

4. $500 - 376 =$

5.
```
  426
x   7
```

6.
```
  307
x  22
```

Helpful Hints	* Sometimes it is necessary to read a problem more than once.
	* Sometimes it is good to circle important words and numbers.
	*Sometimes drawing a sketch or diagram is useful.

S1. A square field has sides 258 ft. How far is it all the way around the field?	1.
	2.
S2. Frank earns 16 dollars per hour. How much does he earn during a 40-hour work week?	3.
	4.
1. Fremont's population is 202,156. Redford's population is 191,216. How much greater is Fremont's population than Redford's?	5.
2. A farmer has 360 cows and decides to sell ½ of them. How many cows will he sell?	Score

3. In a school there are 346 sixth-graders, 408 seventh-graders, and 465 eighth-graders. What is the total number of students?

4. A school has 600 students and needs to divide them into 25 equally sized classes. How many will be in each class?

5. Sam had 356 points on four tests. What was his average score?

Review Exercises

1. A car traveled 440 miles in 8 hours. What was its average speed per hour?

2.
$$624 \\ \times \ 7$$

3. A school has 25 classes with thirty students in each class. How many students are there altogether?

4.
$$467 \\ 29 \\ 736 \\ + \ 27$$

5. Billie earned $4,137 this month and $5,198 last month. How much more did she earn last month?

6.
$$7,115 \\ - \ 678$$

Helpful Hints

The following problems require two steps.

1. Read the problem carefully.
2. Find the important facts and numbers.
3. Decide what operations to use and in what order to use them.
4. Solve the problem.

Label your answer with a word or short phrase.

S1. Sol's test scores were 81, 90, and 99. What was her average score?

S2. A farmer had nine crates of potatoes that weighed 180 pounds each. He also had seven sacks of tomatoes that weighed a total of 365 pounds. What was the total weight of the potatoes and tomatoes?

1. A woman buys a car making a $3,000 down payment and then agrees to make 48 monthly payments of 360 dollars. What is the total cost of the car?

2. Last week a man worked 12 hours a day for five days. This week he worked 57 hours. How many hours did he work in all?

3. A theater has 15 rows of seats with 25 seats in each row. If 19 of the seats are vacant, how many are taken?

4. A tank holds 250,000 gallons of fuel. If 27,600 gallons are removed one day, and 35,750 gallons are removed the next day, how many gallons are left?

5. A car can travel 32 miles per gallon of gas. How many gallons are required to travel 448 miles? If gas is $3 per gallon, how much would the gas cost?

1.
2.
3.
4.
5.
Score

Review Exercises

1.   642
    x 23

2. A library has 47,109 fiction books and 39,783 non-fiction books. What is the total number of books in the library?

3. 30 ⟌ 705

4. If a plane travels 455 miles per hour, how far will it travel in twelve hours?

5. 12 ⟌ 1,976

6. Steve earned $65,650 this year and $58,955 last year. How much more did he earn this year?

Helpful Hints	* Sometimes reading the problem more than once is helpful. * Finding the important facts and numbers and circling them can be helpful. * Decide which operation comes first. * Solve the problem. Label the answer with a word or short phrase.

S1. Sue, Harry, and Linda had a car wash, earning $375 on Friday, $360 on Saturday, and $420 on Sunday. To divide the earning equally, how much would each of them receive?	1.
	2.
S2. Jackie wants to send out 112 party invitations. If the invitations come in boxes of 15, how many boxes must she buy? How many invitations will be left over?	3.
	4.
1. Nine buses hold 85 passengers each. If 693 people buy tickets for a trip, how many seats won't be taken?	5.
	Score

2. Six people are at a party. For refreshments there are three pizzas which are sliced into eight pieces each. How many pieces will each person get?

3. A carpenter bought three hammers for 17 dollars each and two saws for 23 dollars each. What was the total cost?

4. A school has 314 boys and 310 girls. If they are to be grouped into equal classes of 26 each, how many classes will there be?

5. Jim worked 48 weeks last year. Each week he worked 38 hours. If he worked an additional 240 hours of overtime, how many hours did he work in all?

Review Exercises

1. Joel made 540 gift cards. He wants to sell them in boxes of 30. How many boxes will he need?

2.
$$9,176$$
$$- 2,977$$

3. A garden is in the shape of a square. If each side is 29 ft., how far is it all around the garden?

4.
$$208$$
$$\times\ 35$$

5. Erica's test scores were 96, 88, 97, and 95. What was the total of all her scores? What was her average score?

6. $7 \overline{\smash{)}2,375}$

Helpful Hints

* Read the problem carefully.
* Find and circle the important facts and numbers.
* Decide on the correct order of operations.
* After you have solved the problem and labeled it, ask yourself whether the answer makes sense.
* Sometimes it is helpful to make a sketch.

S1. John bought a shirt for $19 and a pair of shoes for $23. If he paid with a $50 bill, how much change would he receive?

S2. A group of hikers set out on an 80-mile hike. If they hiked 9 miles per day for seven days, how many miles would be left to hike?

1. A garden in the shape of a rectangle is 140 feet long and 95 feet wide. What is the distance around the garden?

2. An orchard has eight rows of trees with 15 trees in each row. If each tree produces six bushels of fruit, how many bushels will be produced in all?

3. Bill worked 56 hours this week and 44 hours last week. If he is paid twelve dollars per hour, what were his total earnings?

4. Buses hold 65 people. If 95 parents and 165 students are attending a football game, how many buses will be needed?

5. How old will a man born in 1929 be in 2016?

1.
2.
3.
4.
5.
Score

Review Exercises

1. 41 $\overline{)861}$

2. The enrollment at Wilbur High School is 3,015. Last year the enrollment was 2,965. How much was this year's increase in enrollment?

3. $765 + 97 + 399 =$

4. If five pounds of beef cost $65, what is the price per pound?

5. 605
    x 70

6. Yuri scored a total of 665 points on seven tests. What was her average score?

Helpful Hints

Use what you have learned to solve the following problems.
* Review some of the helpful hints from previous pages.

	1.
S1. Three friends took a camping trip. Food costs were 171 dollars and camping supplies were 273 dollars. If they divided the costs equally, how much would each pay?	2.
	3.
S2. A man bought a car for $15,000. If the down payment was $3,000 and the rest was to be paid in 48 equal monthly payments, how much are the payments?	4.
	5.
1. A school has 180 boys and 160 girls. Tickets to the class picnic are 8 dollars each. If everyone attends, what will be the cost of the tickets?	Score

2. A new car costs $22,000. After one year it lost $2,800 of its value. After two years it lost another $3,350 in value. How much was the car's value after two years?

3. A family took a five-day trip. The first day they drove 385 miles. Each of the next four days they drove 275 miles. How many miles long was the trip?

4. A tank contained 55,000 of fuel. One day 12,500 gallons were added. The next day 15,575 gallons were removed. How many gallons are left in the tank?

5. A family bought a home for $235,000. They spent $75,500 on remodeling it. Then they sold the home for $400,000. What was the profit?

Review Exercises

1. Tom had test scores of 75, 84, and 81. What was his average score?

2.   423
    x 27

3. A school has twelve classes of 32 students each. How many students are there in the school?

4.   15 | 3,015

5. Last year Jose worked 48 weeks and worked forty hours per week. How many hours did he work last year?

6. $9,001 - 279 =$

Helpful Hints

When working with multi-step problems it is important to read very carefully and at least twice.

1. Find the important facts and numbers.
2. Decide which operations to use and in what order.
3. Draw a diagram if necessary.

S1. Jeri bought six dozen hot dogs at three dollars per dozen and three dozen burger patties at seven dollars per dozen. How much did he spend altogether?

S2. A rectangular yard is 16 feet by 24 feet. How many feet of fencing is required to enclose it? If each 5-foot section costs $35, how much will the fencing cost?

1. Two trains leave the station in opposite directions, one at 85 miles per hour, the other at 75 miles per hour. How far apart will they be in 3 hours?

2. A factory can make a table in 130 minutes and a chair in eight minutes. How long will it take to make seven tables and 15 chairs? Express the answer in hours and minutes.   (60 minutes = 1 hour)

3. A farmer had 600 pounds of apples. He gave 200 pounds to neighbors and sold one-half of the remainder at three dollars per pound. How much did he make selling the apples?

4. Tom finished a 30-mile walkathon. He collected pledges of $26 for each of the first 25 miles and $50 for each remaining mile. How much did he collect in all?

5. A carpenter bought three hammers at 17 dollars each, two saws at 27 dollars each, and a drill for 58 dollars. How much did he spend in all?

1.	
2.	
3.	
4.	
5.	
Score	

Reviewing Whole Number Problem Solving

1. Lincoln High School's enrollment is 4,763 and Jefferson High School's enrollment is 4,969. What is the total enrollment for both schools?	1.
	2.
	3.
2. A plane travels 6,600 miles in twelve hours. What is its average speed per hour?	4.
	5.
3. Sophia had test scores of 75, 83, 78, 93, and 96. What was her average score?	6.
	7.
4. A car traveled 265 miles each day for seven days and 325 miles on the eighth day. How many miles did it travel in all?	8.
5. A car traveled 288 miles. It averaged 24 miles per gallon of gas. If gas costs $3 per gallon, how much did the trip cost?	9.
	10.
6. A plumber purchased 3 sinks for $129 each, three bathtubs for $372 each, and five faucets for $23 each. How much did he spend in all?	Score

7. A school has 240 seventh-graders and 300 eighth-graders. If they are to be placed in equally sized classes of 30 students each, how many classes will there be?

8. A woman bought a car for $15,000. She made a $3,000 down payment and paid the rest in equal payments of $400. How many payments did she make?

9. A man wants to put a fence around a square lot with sides 68 feet. How many feet of fencing is needed? If each 8-foot section of fence costs $25, how much will the fence cost?

10. A family is planning a seven-day vacation. Lodging costs 125 dollars per day, food is 85 dollars per day, and entertainment is 150 dollars per day. How much will the trip cost?

To add or subtract fractions with unlike denominators, find the least common denominator. Multiply each fraction by one to make equivalent fractions. Finally, add or subtract.

Examples:

$$\frac{2}{5} \times \frac{2}{2} = \frac{4}{10}$$
$$+ \frac{1}{2} \times \frac{5}{5} = \frac{5}{10}$$
$$\boxed{\frac{9}{10}}$$

$$\frac{5}{6} \times \frac{2}{2} = \frac{10}{12}$$
$$+ \frac{1}{4} \times \frac{3}{3} = \frac{3}{12}$$
$$\frac{13}{12} = \boxed{1\frac{1}{12}}$$

When adding mixed numerals with unlike denominators, first add the fractions. If there is an improper fraction, make it a mixed numeral. Finally, add the sum to the sum of the whole numbers.

*Reduce fractions to lowest terms.

Example:

$$3\frac{2}{3} \times \frac{2}{2} = \frac{4}{6}$$
$$+ 2\frac{1}{2} \times \frac{3}{3} = \frac{3}{6}$$
$$5 \qquad \frac{7}{6} = 1\frac{1}{6} = \boxed{6\frac{1}{6}}$$

To subtract mixed numerators with unlike denominators, first subtract the fractions. If the fractions cannot be subtracted, take one from the whole number, increase the fraction, then subtract.

Examples:

$$^{5}6\frac{1}{6} = \frac{2}{12} + \frac{12}{12} = \frac{14}{12}$$
$$- 3\frac{1}{4} = \frac{3}{12}$$
$$\boxed{2\frac{11}{12}}$$

$$7\frac{1}{2} \times \frac{3}{3} = \frac{3}{6}$$
$$- 2\frac{1}{3} \times \frac{2}{2} = \frac{1}{6}$$
$$5 \qquad \frac{2}{6} = \boxed{5\frac{1}{3}}$$

S1.
$$3\frac{1}{7}$$
$$- 1\frac{5}{7}$$

S2.
$$6\frac{1}{2}$$
$$+ 3\frac{3}{4}$$

1.
$$\frac{8}{9}$$
$$- \frac{1}{2}$$

2.
$$\frac{8}{9}$$
$$- \frac{1}{6}$$

3.
$$7$$
$$- 2\frac{3}{5}$$

4.
$$5\frac{1}{8}$$
$$+ 3\frac{1}{2}$$

5.
$$7\frac{7}{8}$$
$$+ 3\frac{3}{8}$$

6.
$$7\frac{1}{2}$$
$$- 2\frac{3}{4}$$

7.
$$3\frac{4}{5}$$
$$+ 4\frac{2}{3}$$

8.
$$6\frac{1}{2}$$
$$- 3$$

9.
$$\frac{7}{16}$$
$$+ \frac{1}{4}$$

10.
$$4\frac{5}{6}$$
$$+ 3\frac{3}{4}$$

1.	
2.	
3.	
4.	
5.	
6.	
7.	
8.	
9.	
10.	
Score	

When multiplying common fractions, first multiply the numerators. Next, multiply the denominators. If the answer is an improper fraction, change it to a mixed numeral.

Examples:

$$\frac{3}{4} \times \frac{2}{7} = \frac{6}{28} = \boxed{\frac{3}{14}} \qquad \frac{3}{2} \times \frac{7}{8} = \frac{21}{16} = \boxed{1\frac{5}{16}}$$

If the numerator of one fraction and the denominator of another have a common factor, they can be divided out before you multiply the fractions.

Examples:

$$\frac{3}{\overset{1}{\cancel{4}}} \times \frac{\overset{2}{\cancel{8}}}{11} = \boxed{\frac{6}{11}} \qquad \frac{7}{\overset{4}{\cancel{8}}} \times \frac{\overset{3}{\cancel{6}}}{5} = \frac{21}{20} = \boxed{1\frac{1}{20}}$$

When multiplying whole numbers and fractions, write the whole number as a fraction and then multiply.

Examples:

$$\frac{2}{3} \times 15 = \qquad \frac{3}{4} \times 9 =$$

$$\frac{2}{\overset{1}{\cancel{3}}} \times \frac{\overset{5}{\cancel{15}}}{1} = \frac{10}{1} = \boxed{10} \qquad \frac{3}{4} \times \frac{9}{1} = \frac{27}{4} = \boxed{6\frac{3}{4}}$$

To multiply mixed numerals, first change them to improper fractions, then multiply. Express answers in lowest terms.

Example:

$$1\frac{1}{2} \times 1\frac{5}{6} = \frac{\overset{1}{\cancel{3}}}{2} \times \frac{11}{\underset{2}{\cancel{6}}} = \frac{11}{4} = \boxed{2\frac{3}{4}}$$

To divide fractions, find the reciprocal of the second number, then multiply the fractions.

Examples:

$$\frac{2}{3} \div \frac{1}{2} = \qquad\qquad 2\frac{1}{2} \div 1\frac{1}{2} = \frac{5}{2} \div \frac{3}{2} =$$

$$\frac{2}{3} \times \frac{1}{2} = \frac{4}{3} = \boxed{1\frac{1}{3}} \qquad \frac{5}{2} \times \frac{2}{3} = \frac{5}{3} = \boxed{1\frac{2}{3}}$$

S1. $\quad 1\frac{1}{4} \times 2\frac{2}{5} =$

S2. $\quad 5\frac{1}{2} \div 1\frac{1}{2} =$

1. $\quad \dfrac{12}{13} \times \dfrac{3}{24} =$

2. $\quad \dfrac{3}{4} \times 36 =$

3. $\quad \dfrac{7}{8} \times 2\frac{1}{7} =$

4. $\quad 2\frac{1}{3} \times 3\frac{1}{2} =$

5. $\quad \dfrac{3}{4} \div \dfrac{1}{2} =$

6. $\quad 3\frac{1}{2} \div \dfrac{1}{2} =$

7. $\quad 3\frac{2}{3} \div 1\frac{1}{2} =$

8. $\quad 3\frac{3}{4} \div 1\frac{1}{8} =$

9. $\quad 6 \div 2\frac{1}{3} =$

10. $\quad 2\frac{2}{3} \div 2 =$

1.

2.

3.

4.

5.

6.

7.

8.

9.

10.

Score

Review Exercises

1. $\dfrac{1}{2}$
   $+ \dfrac{1}{3}$

2. $1\dfrac{1}{5}$
   $+ 2\dfrac{2}{3}$

3. $2\dfrac{1}{2}$
   $+ 2\dfrac{2}{3}$

4. $3\dfrac{1}{2}$
   $- 1\dfrac{1}{3}$

5. $6\dfrac{1}{3}$
   $- 1\dfrac{1}{5}$

6. 7
   $- 1\dfrac{1}{5}$

Helpful Hints	1. Read the problem carefully. 2. Find and circle the important facts and numbers. 3. Decide what operations to use. 4. Solve the problem and label the answer with a word or phrase.

S1. A baker used $2\dfrac{1}{4}$ cups of flour for a cake and $3\dfrac{1}{2}$ cups for a pie. How much flour did he use in all?

S2. Steve earned 60 dollars and spend two-thirds of it. How much did he spend? (Hint: "of" means multiply.)

1. Bill weighed $124\dfrac{1}{4}$ pounds last year. This year he weighs $132\dfrac{3}{4}$ pounds. How many pounds did he gain?

2. A ribbon is $5\dfrac{1}{2}$ feet long. It was cut into pieces $\dfrac{1}{2}$ foot long. How many pieces were there?

3. It is $2\dfrac{1}{2}$ miles around a race track. How far will a car travel in 12 laps?

4. If it takes a man $12\dfrac{1}{2}$ minutes to drive to work and $16\dfrac{3}{4}$ minutes to drive home, what is his total commute time?

5. What is the perimeter of a flower bed with sides $8\dfrac{1}{2}$ feet?

1.
2.
3.
4.
5.
Score

Review Exercises

1. $\dfrac{1}{2} \times \dfrac{4}{5} =$

2. $\dfrac{1}{3} \times 2\dfrac{1}{2} =$

3. $2\dfrac{1}{2} \times 1\dfrac{1}{5} =$

4. $\dfrac{2}{3} \div \dfrac{1}{2} =$

5. $4\dfrac{1}{2} \div \dfrac{1}{2} =$

6. $3\dfrac{1}{2} \div 2 =$

Helpful Hints

Look for key words:

1. "Total," "sum," "altogether," and "in all" usually mean addition or multiplication.

2. "How much more than," "difference," "what was the increase," and "how much was left" usually mean subtraction.

3. "Of" means multiply. "Divide" and "average" usually mean division.

1.	
2.	
3.	
4.	
5.	
Score	

S1. Tony bought 6 melons, each of which weighed $2\dfrac{1}{2}$ pounds. What was the total weight of the melons?

S2. If a cook had $5\dfrac{1}{4}$ pounds of beef and used $3\dfrac{3}{4}$ pounds, how many pounds were left?

1. Suzette worked $7\dfrac{1}{2}$ hours on Monday and $6\dfrac{3}{4}$ hours on Tuesday. How many hours did she work in all?

2. If a car travels 50 miles per hour, how far will it travel in $2\dfrac{1}{2}$ hours?

3. A factory can produce a tire in $2\dfrac{1}{2}$ minutes. How many tires can it produce in 40 minutes?

4. The Jones' had 36 pounds of beef in their freezer. They used $\dfrac{3}{4}$ of it. How many pounds of beef did they use?

5. Paul decided to study 12 hours for a test. If he has already studied for $7\dfrac{1}{4}$ hours, how much longer will he study for the test?

Review Exercises

1. $\dfrac{3}{5}$
    $+ \dfrac{1}{2}$

2. $\dfrac{7}{8}$
    $- \dfrac{1}{4}$

3. $2\dfrac{2}{3}$
    $+ 3\dfrac{1}{2}$

4. $2\dfrac{1}{3}$
    $+ 3\dfrac{4}{5}$

5. $6\dfrac{3}{5}$
    $- 1\dfrac{1}{2}$

6. $6\dfrac{1}{5}$
    $- 2\dfrac{1}{2}$

Helpful Hints

*Remember:
1. Read the problem carefully.
2. Find the important facts and numbers.
3. Decide what operation to use.
4. Solve and label your answer with a word or phrase.

* Reduce all fractions to lowest terms.

S1. Marc worked for $6\dfrac{1}{2}$ hours. If he is paid 12 dollars per hour, how much did he earn?

S2. A grocer has 30 pounds of tomatoes. If he wants to pack them into $2\dfrac{1}{2}$ - pound packages, how many packages of tomatoes will he have?

1. Mike worked $3\dfrac{3}{4}$ hours on Monday and $2\dfrac{1}{2}$ hours on Tuesday. How many more hours did he work on Monday than on Tuesday?

2. Allie was $60\dfrac{3}{4}$ inches tall. She grew $2\dfrac{1}{2}$ inches. How tall is she now?

3. Phil's reading assignment is $50\dfrac{1}{2}$ pages. If he has read $15\dfrac{1}{3}$ pages, how much more does he have to read?

4. Elena bought 40 hotdogs for the picnic. If each hotdog weighed $\dfrac{1}{3}$ of a pound, what was the total weight of the hot dogs?

5. A family is taking a 360-mile trip. They have driven $\dfrac{3}{4}$ of the distance. How much farther do they have to drive?

1.
2.
3.
4.
5.
Score

Review Exercises

1. $\dfrac{2}{5} \times \dfrac{10}{11} =$

2. $5 \times 2\dfrac{1}{5} =$

3. $2\dfrac{1}{2} \times \dfrac{4}{5} =$

4. $\dfrac{3}{4} \div \dfrac{1}{2} =$

5. $2 \div 1\dfrac{1}{2} =$

6. $2\dfrac{2}{3} \div 1\dfrac{1}{3} =$

Helpful Hints	* Sometimes it's necessary to read a problem more than once. * It's helpful sometimes to circle important words and numbers. * Drawing a diagram can sometimes be useful.

S1. Each side of a square garden is $8\dfrac{1}{4}$ ft.
What is the distance around the garden?

S2. One loaf of bread weighs $1\dfrac{1}{2}$ pounds.
How much do 20 loaves weigh?

1. The distance around a square-shaped window is $20\dfrac{4}{5}$ feet.
What is the length of each side of the window?

2. Mr. Gilbert bought $5\dfrac{1}{2}$ pound of beef. If he divides it into 11
equally sized steaks, how much will each steak weigh?

3. Ronnie is taking a $12\dfrac{1}{2}$ mile hike. If he has gone $8\dfrac{2}{3}$ miles,
how much farther does he need to hike?

4. A garden in the shape of a rectangle is $8\dfrac{1}{2}$ feet by $5\dfrac{1}{4}$ feet.
What is the total distance around the garden?

5. Remy earned $6,300 last month. If $\dfrac{1}{3}$ of this amount goes towards
his house payment, how much is his house payment?

1.
2.
3.
4.
5.
Score

Review Exercises

1. Bill is $60\frac{1}{2}$ inches tall and Sally is $62\frac{1}{4}$ inches tall. How much taller is Sally than Bill?

2. $\frac{3}{4}$
   $+ \frac{1}{3}$
   _____

3. Bill has read $\frac{3}{4}$ of a 28 page assignment. How many pages has he read?

4. $\frac{7}{8}$
   $- \frac{1}{4}$
   _____

5. Vica caught two fish. One weighed $2\frac{3}{4}$ pounds and the other weighed $3\frac{1}{2}$ pounds. What was the total weight of the fish?

6. $2\frac{3}{4}$
   $+ 2\frac{1}{2}$
   _____

Helpful Hints	When working with two-step problems it is necessary to read the problems more carefully. * Decide which operations to use and in which order. * Reduce all fractions to lowest terms.

S1. A tailor had $8\frac{1}{2}$ yards of cloth. He cut off three pieces that were $1\frac{1}{2}$ yards long each. How much of the cloth was left?

S2. A man had 56 dollars. He gave his son $\frac{1}{4}$ of it and his daughter $\frac{1}{2}$ of it. How much did he have left?

1. A painter needs seven gallons of paint. He already has $2\frac{1}{2}$ gallons in one bucket, and $3\frac{1}{4}$ gallons in another. How many more gallons of paint does he need?

2. There are 30 people in a class. If $\frac{2}{5}$ of them are boys, then how many girls are in the class?

3. Eva worked $5\frac{1}{2}$ hours on Saturday and $6\frac{3}{4}$ hours on Sunday. If she was paid 12 dollars per hour, how much did she earn?

4. Michelle made 36 bracelets last week and 28 bracelets this week. If she sold $\frac{3}{8}$ of them, how many bracelets did she sell?

5. Joe's ranch has 4,000 acres. If $\frac{1}{4}$ of the ranch was used for crops and $\frac{2}{3}$ of the remainder was used for grazing, how many acres were for grazing?

1.
2.
3.
4.
5.
Score

Review Exercises

1. $3\frac{1}{2} \div \frac{1}{4} =$

2. How many $\frac{1}{2}$ pound beef patties are there in $5\frac{1}{2}$ pounds of beef?

3. $5 \div 1\frac{1}{2} =$

4. It takes $1\frac{1}{4}$ yards of cloth to cover a chain. How much cloth is needed to cover 8 chairs.

5. $3\frac{3}{4} \div 1\frac{1}{4} =$

6. It took Lola $12\frac{3}{4}$ minutes to drive to work and $10\frac{1}{3}$ minutes to drive home. What was her total commute time?

Helpful Hints

* Sometimes reading the problem more than once is helpful.
* Finding the important facts and numbers and circling them can be helpful.
* Carefully decide the order of operations.
* After solving the problem, the answer should make sense.

S1. A family was going to take a 360 mile trip. They travelled $\frac{1}{2}$ of the total distance the first day and $\frac{1}{3}$ of the total distance the second day. How many more miles were left to travel?

S2. Twenty feet of copper wire is cut into pieces $2\frac{1}{2}$ feet long. If each piece sells for $6, how much would all the pieces cost?

1. A farmer picked $6\frac{1}{2}$ bushels each day for five days. He then sold $15\frac{1}{2}$ bushels. How many bushels were left to sell?

2. A man bought $12\frac{3}{4}$ pounds of beef. He put $10\frac{1}{4}$ pounds in the freezer He used $\frac{3}{5}$ of the rest for cooking a stew. How many pounds did he use for the stew?

3. Land is selling for $30,000 per acre. Mr. Roberts bought $1\frac{1}{2}$ acres on Monday. On Tuesday he decided to buy an additional $2\frac{1}{2}$ acres. How much did he pay in all?

4. Mr. Jones earned $4,000. If $\frac{1}{4}$ of his pay goes towards his car payment and $\frac{2}{5}$ of his pay goes towards his house payment, what is the total cost of his housing and car payment?

5. Stewart needs to read 45 pages for English, 60 pages for history, and 63 pages for science. If he has read $\frac{2}{3}$ of the total amount of pages, how many pages has he read?

1.
2.
3.
4.
5.
Score

Review Exercises

1. $\frac{3}{4}$ of $2\frac{1}{2} =$

2. Ellie earned 60 dollars and deposited $\frac{4}{5}$ of it into her savings account. How much did she deposit?

3. $\begin{array}{r} 736 \\ 47 \\ -516 \\ \hline \end{array}$

4. A car traveled at the speed of 65 miles per hour for 8 hours. How far did the car travel?

5. $30\overline{)639}$

6. Will's weight was $155\frac{1}{4}$ pounds if he increased his weight by $5\frac{2}{3}$ pounds. How much does he weigh now?

Helpful Hints

* Read the problem carefully   * Find and circle the important facts and numbers.
* Decide on the correct order of operations.
* After you have solved the problem and labeled it, ask yourself if the answer makes sense.
* Sometimes drawing a diagram can be helpful.

S1. Al spent $\frac{1}{2}$ of his earnings and deposited $\frac{1}{3}$ of it into his savings account. What fraction of his earnings did he have left?

S2. John needs to read a 256-page novel. He must also read $\frac{3}{5}$ of a 250-page science book. What is the total number of pages that John must read?

1. Lisa rode her bike $6\frac{1}{2}$ miles on Monday, $7\frac{1}{4}$ miles on Tuesday, and $5\frac{3}{4}$ miles on Wednesday. If she wants to ride a total of 25 miles, how much farther does she need to ride?

2. Gwen has 30 ounces of cookie dough. She is going to bake $\frac{3}{4}$-ounce cookies and sell them for 2 dollars each. How many cookies will she sell, and how much will she earn selling them?

3. The Smiths had an eight-pound meatloaf and finished $\frac{3}{4}$ of it. The next day Mr. Smith finished $\frac{1}{4}$ of the leftovers. How many pounds of the original meatloaf was left?

4. Carlos needs to work 40 hours this week. He worked $6\frac{1}{2}$ hours on Monday and $5\frac{3}{4}$ hours on Tuesday. How many more hours does he need to work this week?

5. Sue wants to put trim around a square window with sides $7\frac{1}{2}$ inches. If trim comes in three-inch sections and each section costs two dollars, how much will it cost to trim the window?

1.
2.
3.
4.
5.
Score

Review Exercises

1. $2\frac{1}{2}$
  $+\,3\frac{3}{4}$
  _____

2. $4\frac{1}{5}$
  $-\,3\frac{2}{3}$
  _____

3. $\frac{3}{4}$ of $2\frac{1}{4} =$

4. $5 \times 1\frac{3}{4} =$

5. $\frac{3}{4} \div \frac{1}{3} =$

6. $2\frac{1}{2} \div 1\frac{1}{3} =$

Helpful Hints

Use what you have learned to solve the following problems.
* Review some of the "Helpful Hints" sections from previous pages.

S1. A student bought two binders for $2\frac{1}{2}$ dollars each and four books for $3\frac{1}{2}$ dollars each. What was the total cost?

S2. Henry spent 20 dollars for three books. The first book cost $4\frac{1}{2}$ dollars and the second book cost $5\frac{3}{4}$ dollars. What was the cost of the third book?

1. Ellen has 33 ounces of candy. She is going to put them into $2\frac{3}{4}$ ounce bags and sell them for $3 each. What will her earnings be if she sells all the candy?

2. Simi has $15\frac{3}{4}$ pounds of nuts. She is going to keep $7\frac{1}{4}$ pounds and sell the rest for six dollars a pound. How much will she receive for the nuts?

3. Tom had 60 pounds of vegetables. He sold $\frac{3}{4}$ of them and gave $\frac{2}{5}$ of the remainder to friends. How many pounds did he give to friends?

4. Mr. Jones had 40 dollars. He gave $\frac{3}{5}$ of the money to his three children for their allowances. If they divided the money equally, how much would each child receive?

5. A baker can produce $7\frac{1}{2}$ cakes in an hour. If he works for eight hours and sells each cake for 14 dollars, how much money will he make?

1.
2.
3.
4.
5.
Score

Review Exercises

1. $3\frac{3}{4} \div \frac{1}{4} =$

2. $70\overline{)1,976}$

3. $5 \times 1\frac{1}{2} =$

4. $\begin{array}{r} 427 \\ \times\ 26 \\ \hline \end{array}$

5. $\frac{9}{10} - \frac{2}{3} =$

6. $\begin{array}{r} 7,106 \\ -\ 1,697 \\ \hline \end{array}$

Helpful Hints

When working with multi-step problems, remember to read the problem carefully at least twice. Also, remember your basic steps:
1. Fine the important facts and numbers
2. Decide what operations to use and in what order.
3. Solve the problem and label the answer.
4. The answer should make sense.

S1. Mr. Jones' class has 32 students and Mrs. Jensen's class has 40 students. If $\frac{1}{4}$ of Mr. Jones' students got A's and $\frac{3}{8}$ of Mrs. Jensen's students got A's, how many students got A's?

S2. A developer has two plots of land consisting of 39 acres and 49 acres. How many $\frac{1}{4}$-acre lots does he have? If each lot sold for $5,000, how much did he sell all the lots for?

1. There are 400 boys and 350 girls at Anderson School. Three-quarters of the boys take the bus and $\frac{3}{5}$ of the girls take the bus. How many students in all take the bus?

2. A farm has 2,000 acres. If $\frac{3}{5}$ of the farm was used for crops and $\frac{3}{4}$ of the remainder was used for grazing, how many acres were left?

3. Mr. Rosa earned $3,000 last month. $\frac{1}{3}$ of this went towards his car payment, and $\frac{3}{5}$ of the remainder was for his house payment. How much of his money was left.

4. A man had 200 pounds of beef. He gave $\frac{1}{4}$ of it away to neighbors and kept $\frac{3}{5}$ of the remainder. He sold the rest for $4 per pound. How much did he receive for the sale of the beef?

5. Julio has $7\frac{1}{2}$ acres of land. He then divides the land into $\frac{1}{4}$-acre lots and sells each of them for $5,000. If his original cost for the land was $90,000, what was his profit?

| 1. |
| 2. |
| 3. |
| 4. |
| 5. |
| Score |

Reviewing Fractions Problem Solving

1. John earned 80 dollars and spent $\frac{3}{5}$ of it. How much did he spend?

2. Sylvia weighed $120\frac{1}{2}$ pounds. If she reduced her weight by $3\frac{3}{4}$ pounds, how much does she weigh now?

3. A carpenter had a board that was 20 feet long. If he cut it into $2\frac{1}{2}$-foot pieces, how many pieces would he have?

4. A flight to Chicago took $5\frac{2}{3}$ hours. The return flight only took $4\frac{3}{4}$ hours. What was the total time for both flights?

5. Ben worked $5\frac{1}{2}$ hours on Saturday and $6\frac{3}{4}$ hours on Sunday. If he is paid 12 dollars per hour, what was his pay?

6. Julie earned 45 dollars last week and 96 dollars this week. If she spent $\frac{1}{3}$ of her earnings, how much did she spend?

7. Mary has 30 pounds of candy that she will divide into $1\frac{1}{2}$-pound packages. How many packages will she have? If they sell for $5 each, how much money will she make?

8. In a school with 400 students, $\frac{3}{8}$ of the students take French and $\frac{2}{5}$ of the remainder take Spanish. How many take French? How many take Spanish?

9. Jean baked a pie. If she gave $\frac{1}{2}$ to Steve and $\frac{1}{3}$ to Gina, what fraction of the pie did she have left?

10. Roy picked $7\frac{3}{4}$ bushels of apples Monday and $6\frac{3}{4}$ bushels on Tuesday. If he sold $\frac{1}{2}$ of the apples for $100 per bushel, how much did he earn?

1.
2.
3.
4.
5.
6.
7.
8.
9.
10.
Score

To add decimals, line up the decimal points and add as you would whole numbers. Write the decimal points in the answer. Zeroes may be placed to the right of the decimal.

Example:　Add $3.16 + 2.4 + 12$

$$\begin{array}{r} 3.16 \\ 1.63 \\ +12.00 \\ \hline \boxed{17.56} \end{array}$$

To subtract decimals, line up the decimal points and subtract as you would whole numbers. Write the decimal points in the answer. Zeroes may be placed to the right of the decimal.

Examples:

$3.2 - 1.66 = $
$$\begin{array}{r} {}^{2}\overset{111}{\cancel{3}.\cancel{2}0} \\ -1.66 \\ \hline \boxed{1.54} \end{array}$$

$7 - 1.63 = $
$$\begin{array}{r} {}^{6}\overset{9}{\cancel{7}}.\overset{1}{\cancel{0}}0 \\ -1.63 \\ \hline \boxed{5.37} \end{array}$$

---

* Line up the decimals.　　　* Put decimals in the answer.　　　* Zeroes may be added to the right of the decimal.

1.
2.
3.
4.
5.
6.
7.
8.
9.
10.
Score

S1.　$\begin{array}{r} 3.61 \\ 14.4 \\ +\ \ .37 \\ \hline \end{array}$

S2.　$\begin{array}{r} 7.16 \\ -3.473 \\ \hline \end{array}$

1.　$\begin{array}{r} 7.16 \\ 8.92 \\ +7.634 \\ \hline \end{array}$

2.　$\begin{array}{r} 7.6 \\ -1.43 \\ \hline \end{array}$

3.　$4.36 + 5.7 + 6.24 = $

4.　$17.2 - 8.96 = $

5.　$15 - 12.92 = $

6.　$6.93 + 5 + 7.63 = $

7.　$.9 + .7 + .6 = $

8.　$7.16 - 2.673 = $

9.　$27.16 - 16.764 = $

10.　$7.73 + 2.6 + .37 + 15 = $

Multiplying a Decimal by a Whole Number
Multiply as you would with whole numbers.
Find the number of decimal places and place the
decimal point properly in the product.

Examples:

$$2.32 \leftarrow 2 \text{ places}$$
$$\times \quad 6$$
$$\overline{13.92} \leftarrow 2 \text{ places}$$

$$7.6 \leftarrow 1 \text{ places}$$
$$\times \ 23$$
$$\overline{228}$$
$$1520$$
$$\overline{174.8} \leftarrow 1 \text{ places}$$

Multiplying a Decimal by a Decimal
Multiply as you would with whole numbers.
Find the number of decimal places and place
the decimal point properly in the product.

Examples:

$$2.63 \leftarrow 2 \text{ places}$$
$$\times \ .3 \leftarrow 1 \text{ place}$$
$$\overline{.789} \leftarrow 3 \text{ places}$$

$$.724 \leftarrow 3 \text{ places}$$
$$\times \ .23 \leftarrow 2 \text{ places}$$
$$\overline{2172}$$
$$14480$$
$$\overline{.16652} \leftarrow 5 \text{ places}$$

Dividing a Decimal by a Whole Number
Divide as you would with whole numbers.
Place the decimal point directly up.

Examples:

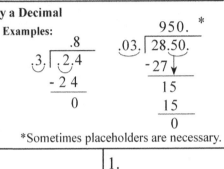

Dividing a Decimal by a Decimal
Move the decimal point in the divisor the number of places necessary to make it a whole number. Move the decimal point in the dividend the same number of places.

*Sometimes placeholders are necessary.

	1.
S1. 3.24 $\times$ 2.3	2.
S2. .5 $\overline{).325}$ 1. 4.26 $\times$ 3	3.
	4.
2. 3.4 $\times$ 16 3. 3.08 $\times$ 1.6 4. 6.32 $\times$ 23.4	5.
	6.
	7.
5. 2 $\overline{)2.68}$ 6. 5 $\overline{)7.3}$ 7. .003 $\overline{)1.2}$	8.
	9.
	10.
8. .5 $\overline{).375}$ 9. .15 $\overline{).0075}$ 10. 8.7 $\overline{).1131}$	Score

Review Exercises

1. 3.56
    2.723
    + 4.96
    _____

2. $3.16 + 15 + 2.79 =$

3. $7.13 - 2.652 =$

4.   6.2
    $-$ 3.564
    _____

5.   2.14
    x   3
    _____

6.   .58
    x 32
    _____

| **Helpful Hints** | 1. Read the problem carefully.
2. Find the important facts and numbers.
3. Decide which operation to use.
4. Solve the problem and label it with a work or short phrase. | * Be careful with decimal placement. |

S1. If calculators cost $7.95 each, what is the cost of eight calculators?

S2. A jet traveled 1,250 miles in 2.5 hours. What was its average speed?

1. Potatoes cost $3.23 a pound and carrots cost $2.89 a pound.
    How much more do potatoes cost per pound?

2. A man bought a desk for $375.50, a chair for $119.90, and a lamp
    for $23.45. What was the total cost of the items?

3. A square lot has sides 48.5 feet long. How far is it around the lot?

4. 12 cans of corn cost $13.68. What is the cost of one can?

5. A baseball glove was on sale for $32.65. If the regular price was
    $45.25, how much can be saved buying it on sale?

1.	
2.	
3.	
4.	
5.	
Score	

Review Exercises

1. .72
    x   .3

2. 2.62
    x .03

3. .003
    x .002

4. 2 ⟌ 3.68

5. 5 ⟌ .13

6. 2 ⟌ .014

Helpful Hints

Look for key words:

1. "Total," "sum," "altogether," and "in all" usually mean addition or multiplication.

2. "How much more than," "difference," "what was the increase," and "how much was left" usually mean subtraction.

3. "Divide" and "average" usually mean division.

1.
2.
3.
4.
5.
Score

S1. A plane can travel 450 miles in one hour. At this rate, how far can it travel in .8 hours?

S2. A sack of potatoes was $3.15. If the price was $.45 per pound, how many pounds were in the sack?

1. Tom weighs 135.6 pounds and Jerry weighs 142.75 pounds. What is their total weight?

2. If 8 pounds of butter costs $7.12, what is the price per pound?

3. An engine uses 3.5 gallons of gas per hour. How many gallons will the engine use in 3.2 hours?

4. If cans of soda cost $.25, how many cans of soda can be bought with $5.00?

5. Steak costs $4.80 per pound. How much will .7 pounds cost?

Review Exercises

1. $.2\overline{)1.34}$

2. $.02\overline{).13}$

3. $.03\overline{)5.1}$

4. $.06\overline{).324}$

5. $.12\overline{)1.104}$

6. $.18\overline{).576}$

Helpful Hints

Sometimes it is necessary to read a problem at least twice. Then decide the necessary operation and solve the problem.

* Be careful with decimal placement.
* Think about your answer and be sure it makes sense.

S1. A man bought groceries that cost a total of $34.16. If he paid with a 50-dollar bill, how much change did he receive?

S2. John bought six boxes of chocolates for $7.95 each. What was the total cost?

1. Twelve pounds of apples cost $27.00. What is the price per pound?

2. Lonnie bought a car for $6545.25 and resold it for $8476.49. What was his profit?

1.
2.
3.
4.
5.
Score

3. Four friends earned $301.92. If they decided to divide the money equally, how much will each person receive?

4. Ken bought a car for $3006.00. If he pays for it in twelve equal payments, how much is each payment?

5. How much will it cost for twenty-four $.41 stamps?

Review Exercises

1. 7.36
    .95
    + 4.93

2. 6 − 2.713 =

3. 6 x .427 =

4. $2.75
    x   8

5. 5 ⟌ $15.75

6. .02 ⟌ 5

Helpful Hints

* When reading a problem, sometimes it is helpful to circle important facts and numbers.

* When the problem has been solved, go back and be sure the answer makes sense.

* Sometimes drawing a diagram is helpful.

	1.
S1. Samuel earns $15.50 per hour. How much does he earn in 40 hours?	2.
S2. A rectangle has a length of 12.5 inches and a width of 8.3 inches. What is the perimeter of the rectangle?	3.
1. A man has $481.24 in his savings account. If he makes a deposit of $242.35, what will his new balance be?	4.
	5.
2. The sale price of a dress is $14.50 less than the regular price. If the regular price is $49.65, what is the sale price?	Score

3. Rainfall for the last four days was 2.12 inches, 3.89 inches, 2.73 inches, and 4.79 inches. What was the total rainfall?

4. A man bought 12.5 gallons of gas for $40.00. What was the price per gallon?

5. 1.5 pounds of beef cost $5.10. What is the price per pound?

Review Exercises

1. $5\overline{)3}$

2. Brian earned 40 dollars and spent $\frac{3}{4}$ of it. How much did he spend?

3. $\begin{array}{r} \frac{3}{5} \\ + \frac{1}{3} \\ \hline \end{array}$

4. A $7\frac{1}{2}$ foot board is cut into 3 equal pieces. How long is each piece?

5. $\begin{array}{r} 3\frac{1}{7} \\ - 1\frac{1}{2} \\ \hline \end{array}$

6. What is the perimeter of a square-shaped window with sides $7\frac{1}{2}$ inches?

Helpful Hints	When working with 2-step problems it is necessary to read the problems more carefully. * Decide which operations to use and in which order. * Be carefully with decimal placement. * Be sure your answer makes sense.

S1. A man bought five bags of chips at $.89 each and a pizza for $8.95. How much did he spend?

S2. Yuri bought a hammer for $6.79 and a screwdriver for $4.75. If he paid with a 20-dollar bill, what was his change?

1. Zach is taking a trip of 192 miles. If his car gets 24 miles per gallon of gas, and gas costs $3.10 per gallon, what is the cost of the trip?

2. Mark worked 40 hours and was paid $12.50 per hour. He received a bonus of $125.75 for overtime. What were his total earnings?

3. Cans of corn are two for $1.19. What is the cost for 12 cans?

4. Jeans are on sale for $12.95. If the regular price is $15.50, how much would be saved by buying two pairs of jeans on sale?

5. Three friends earned $3.65 on Monday, $7.75 on Tuesday, and $9.75 on Wednesday. If they divided the money equally, how much would each of them receive?

1.
2.
3.
4.
5.
Score

Review Exercises

1. A car traveled 365 miles per day for five days. What was the total distance traveled.

2. $2 \div 1\frac{3}{4} =$

3. A factory can produce 34 engines per hour. How many can it produce in 12 hours?

4. $4 \times 2\frac{1}{3} =$

5. A rope $17\frac{1}{2}$ feet long is divided into 7 equal pieces. How long is each piece?

6. $2\frac{1}{2} \div 1\frac{1}{4} =$

Helpful Hints	Ask yourself these questions: 1. What does the problem ask you to do? 2. Are there helpful "keywords"? 3. What are the facts and numbers that will help in finding the answer?

S1. A jacket costs $12.00. Sales tax is .07 of this price. What will be the total price including sales tax? (Hint: "of" often means multiply.)

S2. A man bought 12 gallons of gas at $2.75 per gallon and a quart of oil for $4.79. What was the total cost?

1. Cans of peas are 3 for $1.29. How much would one can of peas and a bag of chips priced at $2.19 cost altogether?

2. Tom's times for the 100-yard dash were 11.8, 12.2, and 12.3. What was his average time?

3. If two pounds of chicken cost $3.60, how much will ten pounds cost?

4. John bought a shirt for $13.55 and a pair of shoes for $27.50. If he gave the clerk a $50 bill, how much change will he receive?

5. Marie is buying a bike. She makes a down payment of $75.00 and pays the rest in 12 monthly installments of $32.75 each. How much does she pay in all?

1.
2.
3.
4.
5.
Score

Review Exercises

1. A car traveled 265 miles in five hours. What was its average speed?

2. $3\overline{\smash{)}7,968}$

3. Last year's attendance at a concert was 13,768. This year the attendance was 17,272. What was the increase in attendance?

4.   307
     x 26

5. It is $1\frac{1}{2}$ miles around a track. What is the distance traveled in 10 laps?

6. $376 + 39 + 778 =$

Helpful Hints

Remember these important steps:

1. Read the problem carefully.
2. Find the important facts and numbers.
3. Decide what operations are necessary and the order in which to use them.
4. Solve the problem and label it with a word or short phrase.

S1. A number is the sum of 12.6 and 14.78, decreased by 2.36. Find the number.

S2. A number is the product of 6.2 and 3.4, minus 1.76. Find the number.

1. A number is the quotient of 15.05 and 7, plus 3.16. Find the number.

2. A number is the difference of 17.2 and 9.16, multiplied by 2.1. Find the number.

1.
2.
3.
4.
5.
Score

3. Tony worked 8 hours per day for 6 days. If he was paid $10.50 per hour, what were his earnings?

4. A group of five friends go to a restaurant. The bill comes to $65.75 plus a tip of $9.00. If they decide to split the cost evenly, how much will each of them pay?

5. Bill bought 2.5 pounds of beef priced at $3.50 per pound. What will be his change from a $10.00 bill?

Review Exercises

1. 3.26 + 4 + 3.96 =

2. 7 − 2.165 =

3. 3 x 7.096 =

4. 5 ⟌ 1.3

5. .3 ⟌ .015

6. .003 ⟌ 15

Helpful Hints

Use what you have learned to solve the following problems.
* "Of" often means multiply.
* Sometimes drawing a diagram is helpful.

1.
2.
3.
4.
5.
Score

S1. A runner ran 7.8 miles on Monday, 8.4 miles on Tuesday, and 7.5 miles on Wednesday. What was his average distance?

S2. A rancher decided to sell .2 of his 4,000 acre ranch. If he sold the land for $7,000 per acre, how much was the sale price?

1. The regular price of a pen is $4.75 and the sale price is $3.85. How much can be saved buying 50 pens on sale?

2. Beef is $4.50 per pound and chicken is $2.50 per pound. What is the cost of .6 pounds of beef and .8 pounds of chicken?

3. Alex worked 30 hours, earning $8.15 per hour. If he bought a video game for $97.50, how much of his earnings was left?

4. Jane wants to trim a painting with a length of 9,5 inches and a width of 7.5 inches. If trim costs $5.00 per inch, what will be the cost?

5. Tom bought 6 CD's for $5.50 each and a DVD for $11.79. What was the total cost?

Review Exercises

1. $2\frac{3}{4} \div \frac{1}{2} =$

2. $2\frac{3}{4} \times \frac{1}{2} =$

3. $7\frac{1}{4}$
   $-3\frac{2}{3}$

4. $5\frac{1}{5}$
   $+3\frac{7}{8}$

5. 7
   $-2\frac{1}{4}$

6. $7\frac{1}{2}$
   $-2\frac{3}{4}$

Helpful Hints	When working with multi-step problems, remember to read the problem carefully at least twice to fully understand what is being asked. * Circling key words and numbers can be quite helpful. * Make sure the answer makes sense. * "Of" often means multiply.

	1.
S1. A woman bought 6 dozen hotdogs at $3.39 per dozen and three dozen burger patties at $6.15 per dozen. How much did she spend in all?	2.
S2. A rectangular yard is 24 feet by 16 feet. How many feet of fence is needed to enclose it? If each 2.5-foot section costs $30.00, how much will the fencing cost?	3.
	4.
1. Mr. Arnold's class has 40 students and Ms. Grey's class has 35 students. If .6 of Mr. Arnold's students got A's and .4 of Ms. Grey's students got A's, how many students got A's altogether?	5.
	Score

2. A carpenter bought 3 hammers for $7.99 each, 2 saws for $14.50 each, and a drill for $26.95. What was the total cost?

3. A farm had 2,000 acres. If .6 of the land was used for crops and .4 of the remainder was used for grazing, how many acres were left?

4. Two cars leave a garage in opposite directions, one at 50.5 miles per hour and the other at 30.25 miles per hour. How far apart will they be in 2 hours?

5. Beans are three cans for $.69 and corn is three cans for $1.23. How much will it cost for two cans of each?

Reviewing Decimal Problem Solving

1. A plane can travel 560 miles in one hour. At this rate, how far can it travel in .8 hours?

2. If seven pounds of butter cost $8.33, what is the price per pound?

3. Arnie weighed 139.5 pounds last year, and this year he weighs 147.3 pounds. How much did his weight increase?

4. A woman earned $127.50 on Monday, $133.79 on Tuesday, and $127.65 on Wednesday. What were her total earnings?

5. If potatoes cost $.55 per pound, how many pounds can be bought with $4.95?

6. On a 308-mile trip, a car averaged 22 miles for each gallon of gas. If gas costs $3.15 per gallon, how much did the trip cost?

7. A student bought five pens for $2.19 each and a binder for $8.39. What was the total cost?

8. Beef is $3.60 per pound and chicken is $3.20 per pound. What is the total cost for 2.5 pounds of beef and 1.5 pounds of chicken?

9. A ranch had 5,000 acres. The owner sold .3 acres of the ranch for $7,000 per acre. How much did he receive for the sale of the land?

10. A woman bought three chairs for $21.95 each and a table for $26.50. How much change did she receive if she paid with a $100.00 bill?

| 1. |
| 2. |
| 3. |
| 4. |
| 5. |
| 6. |
| 7. |
| 8. |
| 9. |
| 10. |
| Score |

Percent means "per hundred" or "hundredths." If a fractions is expressed as hundredths, it can easily be written as a percent.

Examples:

$$\frac{7}{100} = 7\% \qquad \frac{3}{10} = 30\% \qquad \frac{19}{100} = 19\%$$

"Hundredths" = percent
Decimals can easily be changed to percents.

Examples: $.27 = 27\%$ $.9 = .90 = 90\%$

* Move the decimal twice to the right and add a percent symbol.

To change a fraction to a percent, first change the fraction to a decimal, then change the decimal to a percent. Move the decimal twice to the right and add a percent symbol.

Examples:

$$\frac{3}{4} \qquad 4\overline{)3.00} \quad .75 = \boxed{75\%}$$
$$\qquad\qquad -2.8$$

$$\frac{16}{20} = \frac{4}{5} \qquad 5\overline{)4.00} \quad .80 = \boxed{80\%}$$
$$\qquad\qquad\qquad -4.0$$
$$\qquad\qquad\qquad\quad 0$$

Percents can be expressed as decimals and as fractions. The fraction form may sometimes be reduced to its lowest terms.

Examples:

$$.25\% = .25 = \frac{25}{100} = \frac{1}{4} \qquad 8\% = .08 = \frac{8}{100} = \frac{2}{25}$$

S1. $20\% = .$ ___ = ___	S2. $9\% = .$ ___ = ___
1. $16\% = .$ ___ = ___	2. $6\% = .$ ___ = ___
3. $75\% = .$ ___ = ___	4. $40\% = .$ ___ = ___
5. $1\% = .$ ___ = ___	6. $45\% = .$ ___ = ___
7. $12\% = .$ ___ = ___	8. $5\% = .$ ___ = ___
9. $50\% = .$ ___ = ___	10. $13\% = .$ ___ = ___

1.

2.

3.

4.

5.

6.

7.

8.

9.

10.

Score

Finding the Percent of a Number
To find the percent of a number, you may use either fractions or decimals. Use what is the most convenient.

Examples:

Find 25% of 60
.25 x 60

$$\begin{array}{r} 60 \\ \times\ .25 \\ \hline 300 \\ 120 \\ \hline \boxed{15.00} \end{array}$$

OR

$$\frac{25}{100} = \frac{1}{4}$$

$$\frac{1}{\cancel{4}_1} \times \frac{\cancel{60}^{15}}{1} = \frac{15}{1} = \boxed{15}$$

Finding the Percent When finding the percent, first write a fraction, change the fraction to a decimal, then change the decimal to a percent.

Examples:

4 is what percent of 16?

$$\frac{4}{16} = \frac{1}{4}$$

$$\begin{array}{r} .25 = \boxed{25\%} \\ 4\ \overline{)\ 1.00} \\ -\ 8 \\ \hline 20 \\ -\ 20 \\ \hline 0 \end{array}$$

5 is what percent of 25?

$$\frac{5}{25} = \frac{1}{5}$$

$$\begin{array}{r} .20 = \boxed{20\%} \\ 5\ \overline{)\ 1.00} \\ -\ 1.0 \\ \hline 00 \end{array}$$

Finding the Whole
To find the whole when the part and the percent are known, simply change the equal sign "=" to the division sign "÷".

Examples:

6 = 25% of what number
6 ÷ 25% change = to ÷
6 ÷ .25 change % to decimal

$$.25\overline{)\ 6.00}\quad \boxed{24.}$$

12 is 80% of what?
12 ÷ 80%
12 ÷ .8

$$.8\overline{)\ 12.0}\quad \boxed{15.}$$

* Be careful to move decimal points carefully.

Solve the problems.

S1.   4 is what % of 20

S2.   3 = 15% of what?

1.   Find 20% of 210.

2.   Find 6% of 350.

3.   15 is what % of 60?

4.   5 is 20% of what?

5.   15 = 75% of what?

6.   30% of 200 =

7.   18 is what % of 24?

8.   Find 25% of 64.

9.   3 is 5% of what?

10.   16 is what % of 80?

1.

2.

3.

4.

5.

6.

7.

8.

9.

10.

Score

51

Review Exercises

1. 723
    x .6

2. 39.7
    x .06
       2.

3. Find 12% of 60.

4. Find 25% of 70.

5. $\frac{3}{5}$ x 25 =

6. $\frac{1}{4}$ x 20 =

Helpful Hints

When finding the percent of a number in a word problem, you can change the percent of a fraction or a decimal. Always express your answer in a short phrase or sentence.

Example:
A team played 60 games and won 75% of them.
How many games did they win?
Find 75% of 60
.75 x 60

    60
  x .75
  ─────
   300
   420
  ─────
  (45.00)

OR

$\frac{75}{100} = \frac{3}{4}$

$\frac{3}{4_1} \times \frac{60^{15}}{1} = \frac{45}{1} = \boxed{45}$

Answer:
The team won
45 games.

S1. Gloria took a test with 40 problems on it. If he got 80% of the problems correct, how many problems did he get correct?

S2. If 6% of the 500 students enrolled in a school are absent, how many students are present?

1. Marty wants to buy a car that costs $9,000. If he has saved 20% of this amount, how much has he saved?

2. Erin has a stamp collection consisting of 30 stamps. If 70% of the stamps are from the USA, how many stamps are from other countries?

3. A house priced at $150,000 requires a 20% down payment. How much is the down payment?

4. A coat is priced at $60. If the sales tax is 7% of the price, how much is the sales tax? What is the total cost including sales tax?

5. If a car costs 15,000 and loses 20% of its value in one year, how much will the car be worth in a year?

1.	
2.	
3.	
4.	
5.	
Score	

Review Exercises

1.   Find 6% of 80.

2.   Find 60% of 80.

3.   Change $\frac{3}{5}$ to a percent.

4.   3 is what % of 5?

5.   15 is what % of 20?

6.   .9 is what % ?

Helpful Hints

Use what you have learned to solve the following problems.

   * Before solving the problem, change the percent to a fraction or a decimal.

1.	
2.	
3.	
4.	
5.	
Score	

S1.   A ranch is 5,000 acres. If 70% of the ranch is used for crops, how many acres are used for other purposes?

S2.   A computer priced at $350 is on sale for 15% off. What is the sale price of the computer?

1.   A school has 300 students. If 40% of the students are boys, how many girls are there in the school?

2.   Gordon bought a CD for $16. If the sales tax is 8%, what it the total cost of the CD?

3.   Bill had a 90-page reading assignment. If he decided to read 60% of the pages before he eats dinner, how many pages will he have left to read after dinner?

4.   A book was priced at $7.50. If the price was reduced by 20%, what is the new price?

5.   A team played 40 games and won 80% of them. How many games did they win?

Review Exercises

1. Change $\frac{2}{5}$ to a percent.

2. Change $\frac{21}{28}$ to a percent.

3. 5 is what % of 20?

4. 18 is what % of 20?

5. Find 30% of 80.

6. Find 12% of 80.

Helpful Hints

When finding the percent first write a fraction, change the fraction to a decimal, then change the decimal to a percent.

Example:
A team played 20 games and won 15 of them.
What percent of the games did they win?

15 is what % of 20?

$$\frac{15}{20} = \frac{3}{4}$$

$$
\begin{array}{r}
.75 = \boxed{75\%} \\
4\,\overline{\smash)3.00} \\
-28 \\
\hline
20 \\
-20 \\
\hline
0
\end{array}
$$

Answer:
They won 75% of the games.

S1. On a test with 25 questions, Felicia got 20 correct.
     What percent did she get correct?

S2. In a class of 25 students, 15 are girls.
     What percent are boys?

1. A worker earned $200 and spent $150 of the money.
    What percent of the earnings did he spend?

2. If 21/25 of a class were present at school,
    what percent of the class was present?

3. A pitcher threw 12 pitches and nine of them were strikes.
    What percent were strikes?

4. On a test with 28 questions, Anna got 21 of them correct.
    What percent of the questions did she miss?

5. Twelve of the 20 students in a class rode the bus to school.
    How many students ride the bus?

1.
2.
3.
4.
5.
Score

54

Review Exercises

1. Find 6% of 90.

2. 12 is what % of 15?

3. 3 = 20% of what?

4. Find 40% of 65.

5. Change $\frac{15}{50}$ to a percent.

6. 6 = 25% of what?

Helpful Hints

Use what you have learned to solve the following problems.

* When solving the problem be sure to reduce fractions to lowest terms.

	1.
S1. Peter has finished 18 of the 24 questions on a test. What percent of the test has he finished?	2.
	3.
S2. Sally earned $25. She saved $15 and spent $10. What percent of her earnings did she spend?	4.
1. A team played 12 games this month and 13 games last month. If they won a total of 20 games, what percent of the games played did they win?	5.
	Score

2. Forty players tried out for a team and only 12 made the team. What percent of those who tried out made the team?

3. Twenty-seven is what percent of 36?

4. Jill earned $60. She spent $15 on a calculator and $9 on a binder. What percent of her earnings did she spend?

5. Three-fifths of a class received A's. What percent of the class received A's?

Review Exercises

1. 3 is 20% of what?

2. 40 is 25% of what?

3. Find 6% of 200.

4. Find 60% of 200.

5. 3 is what % of 15?

6. 45 is what % of 50?

Helpful Hints

When finding the whole, simply change the equal sign to division. **Examples:**

5 people got A's on a test.
This is 20% of the class.
How many are in the class?

5 = 20% of what?
5 ÷ 25%
5 ÷ .2

$$.2\overline{)5.0}$$ = 25.

There are 25 in the class.

200 students at a school are 7th graders.
If this is 25% of the students, how many students are there in the school?

200 = 25% of what?
200 ÷ 25%
200 ÷ .25

$$.25\overline{)200.00}$$ = 800.

There are 800 students in the school.

S1. A team won five games. If this is 20% of the total games played, how many games have they played?

S2. Bill has 24 USA stamps in his collection. If this is 20% of his collection, how many stamps does he have?
How many are not USA stamps?

1. A man spent six dollars, which was 20% of his earnings. How much were his earnings?

2. Seven is 20% of what number?

3. A farmer decided to sell 50 bushels of corn. If this is 5% of his corn harvest, how many bushels were in his harvest?

4. A basketball player made nine shots. This was 75% of the shots taken. How many shots did he take? How many shots did he miss?

5. Sonya got 24 problems correct on a test. Her score was 80%. How many problems were on the test?

1.
2.
3.
4.
5.
Score

Review Exercises

1. 3 is 15 % of what?

2. 2 is what percent of 5?

3. Find 90% of 250.

4. 15 is 60% of what?

5. 28 is what % of 35?

6. Find 9% of 250.

Helpful Hints	Use what you have learned to solve the following problems. * Be careful with decimal placement.

S1. If you get a score of 70% and you got 35 questions correct, how many questions are on the test?

S2. Twenty people passed a test. This was 16% of those who took the test. How many took the test?

1. There are eight red marbles in a bag. If this is 40% of the marbles in the bag, how many marbles are in the bag?
    How many are not used?

2. Mr. Garcia paid $5,000 in taxes. If this was 20% of his earnings, how much were his earnings?

3. Six is 40% of what?

4. Fifteen students in a class received an award. How many students were in the class if this was 30% of the class?

5. A team won 12 games. If this was 75% of the games played, how many games did they lose?

1.
2.
3.
4.
5.
Score

Reviewing Percent Problem Solving

1. Find 25% of 220

2. 6 is 20% of what?

3. 8 is what % of 40?

4. Change $\frac{19}{20}$ to a percent.

5. In a class of 40 people, 20% received A's. How many received A's?

6. A team played 24 games and won 18 of them.
   What percent of the games did they win?

7. Seven students qualified to take advanced French. If this was 20%
   of the class, how many students are in the class?

8. If $\frac{12}{16}$ of the students in a school walk to school,
   what percent of the students walk to school?

9. A ranch has 6,000 acres. If 60% of the ranch is used for crops,
   how many acres are used for other purposes?

10. In a class of 30 students, 6 of them received A's. What percent did
    not receive A's?

1.
2.
3.
4.
5.
6.
7.
8.
9.
10.
Score

58

More Reviewing Percent Problem Solving

1. Change $\frac{12}{20}$ to a percent.

2. 6 is 15% of what number?

3. 8 is what percent of 32?

4. Find 15% of 450.

5. A team lost five games. If this was 20% of all the games they played, how many games did they play?

6. A school has 450 students. If 60% of the students take the bus to school, how many students do not take the bus to school?

7. In a school of 800 students, 600 of them plan to attend summer school. What percent are planning to attend summer school?

8. Seventy percent of the 80 fish in a tank are goldfish. How many goldfish are in the tank?

9. In a class of 40 students, five are taking French and 25 are taking Spanish. What percent of the class is taking a foreign language?

10. A CD is priced at $32. If the sales tax is 8%, what will be the total cost of the CD?

1.
2.
3.
4.
5.
6.
7.
8.
9.
10.
Score

Final Review - Whole Numbers

1. 347
    + 467

2. 614
    723
    17
    + 824

3. 6,403 + 763 + 16,799 =

4. 6,502 + 2,134 + 654 + 24 =

5. 6,093 + 748 + 83 + 769 =

6. 927
    - 648

7. 5,392
    - 1,764

8. 6,053 - 4,639 =

9. 5,000 - 3,286 =

10. 6,003 - 719 =

11. 73
     × 4

12. 7,136
     ×   4

13. 45
     × 37

14. 342
     × 46

15. 643
     × 246

16. 4 ⟌526

17. 4 ⟌1376

18. 40 ⟌568

19. 30 ⟌7614

20. 18 ⟌1243

1.
2.
3.
4.
5.
6.
7.
8.
9.
10.
11.
12.
13.
14.
15.
16.
17.
18.
19.
20.
Score

Final Review - Fractions

1. $\dfrac{4}{7}$
   $+\ \dfrac{1}{7}$

2. $\dfrac{7}{8}$
   $+\ \dfrac{3}{8}$

3. $\dfrac{3}{5}$
   $+\ \dfrac{1}{3}$

4. $3\dfrac{1}{2}$
   $+\ 2\dfrac{3}{8}$

5. $7\dfrac{3}{5}$
   $+\ 6\dfrac{7}{10}$

6. $\dfrac{7}{8}$
   $-\ \dfrac{1}{8}$

7. $6\dfrac{1}{4}$
   $-\ 2\dfrac{3}{4}$

8. 5
   $-\ 2\dfrac{1}{7}$

9. $7\dfrac{3}{5}$
   $-\ 2\dfrac{1}{2}$

10. $7\dfrac{1}{4}$
    $-\ 2\dfrac{1}{3}$

11. $\dfrac{3}{5} \times \dfrac{2}{7} =$

12. $\dfrac{3}{20} \times \dfrac{5}{11} =$

13. $\dfrac{5}{6} \times 24 =$

14. $\dfrac{5}{8} \times 3\dfrac{1}{5} =$

15. $2\dfrac{1}{2} \times 3\dfrac{1}{2}$

16. $\dfrac{5}{6} \div \dfrac{1}{3} =$

17. $2\dfrac{1}{3} \div \dfrac{1}{2} =$

18. $2\dfrac{2}{3} \div 2 =$

19. $5\dfrac{1}{2} \div 1\dfrac{1}{2} =$

20. $6 \div 1\dfrac{1}{3} =$

1.	
2.	
3.	
4.	
5.	
6.	
7.	
8.	
9.	
10.	
11.	
12.	
13.	
14.	
15.	
16.	
17.	
18.	
19.	
20.	
Score	

Final Review - Decimals

1. 4.67
    3.5
  + 3.743

2. .6 + 7.62 + 6.3 =

3. 16.8 + 6 + 7.9 =

4. 36.4
  - 17.8

5. 6.3
  - 3.69

6. 72 - 1.68 =

7. 2.64
  ×   3

8. 2.6
  × 73

9. .63
  × 2.4

10. .126
   × 4.23

11. 10 × 3.65 =

12. 1,000 × 3.6 =

13. $2\overline{)4.64}$

14. $5\overline{)6.7}$

15. $.4\overline{).124}$

16. $.004\overline{)1.2}$

17. $.15\overline{).0045}$

18. $.67\overline{)8.71}$

19. Change $\frac{3}{5}$ to a decimal.

20. Change $\frac{7}{20}$ to a decimal.

1.	
2.	
3.	
4.	
5.	
6.	
7.	
8.	
9.	
10.	
11.	
12.	
13.	
14.	
15.	
16.	
17.	
18.	
19.	
20.	
Score	

Final Review - Graphs

1.

2.

3.

4.

5.

6.

7.

8.

9.

10.

11.

12.

13.

14.

15.

16.

17.

18.

19.

20.

Score

Height of Waterfalls

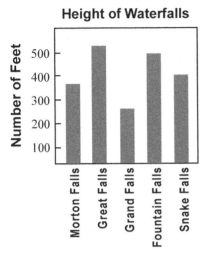

1. Which waterfall is the smallest?

2. Approximately how high is Great Falls?

3. Approximately how much higher is Morton Falls than Grand Falls?

4. Which waterfall is about the same height as Morton Falls?

5. Which waterfall is the fourth highest?

Average Monthly Temperatures

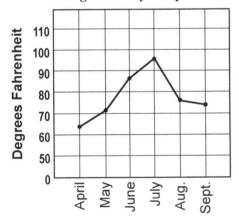

11. What is the average temperature of May?

12. How much cooler was April than July?

13. Which was the second hottest month?

14. Which month's temperature dropped the most from the previous month?

15. What is the difference in temperature between the hottest month and the second hottest month?

Family Budget: $3,000 per month

6. What percent of the family's money is spent on clothing?

7. What percent of the budget is spent on housing?

8. How many dollars are spent on housing per month? (Hint: Find 20% of $3,000.)

9. How many dollars do they save per month?

10. What percent of the family budget is left after paying food, housing, and car expenses?

Fish caught in Drakes Bay in 1991
Salmon
Perch
Cod
Bass
Snapper
Tuna
Each symbol represents 10,000 fish

16. How many perch were caught in 1991?

17. How many more snapper were caught than bass?

18. How many salmon and cod were caught?

19. If the average perch weighs three pounds, how many pounds of perch were caught in 1991?

20. What are the three most commonly caught types of fish?

Final Review - Problem Solving

	1.
	2.
	3.
	4.
	5.
	6.
	7.
	8.
	9.
	10.
	Score

1. A man earned $1,496 last week and $2,018 this week. What were his total earnings?

2. A plane travelled 3,015 miles in five hours. What was the average speed per hour?

3. A student took a test with 60 problems and got $\frac{5}{6}$ of them correct. How many problems did he get correct?

4. Sixteen pounds of nuts were put into bags which each held $1\frac{1}{3}$ pounds. How many bags were there?

5. Alicia earned $16\frac{1}{2}$ dollars on Monday and $12\frac{3}{4}$ dollars on Tuesday. How much more did she make on Monday?

6. If a train can travel 260 miles in one hour, how far can it travel in .8 hours?

7. Hats were on sale for $23.50. If the regular price was $30.25, how much would be saved buying three hats on sale?

8. Stan weighed 96 pounds, Cheryl weighed 120 pounds, and Renee weighed 93 pounds. What was their average weight?

9. Tony earned $65.50 each day for five days and $15.25 on the sixth day. How much did he earn in all?

10. If eight pounds of apples cost $22.48, how much does one pound cost?

11.
12.
13.
14.
15.
16.
17.
18.
19.
20.
Score

11. Antonia took a test with 32 questions. If he missed $\frac{1}{8}$ of them, how many questions did he get correct?

12. A car travelled 348 miles and averaged 29 miles per gallon of gas. If gas was $2.50 per gallon, how much did the trip cost?

13. To tune up his car, Stan bought six spark plugs for $1.95 each and a filter for $9.95. What was the total cost?

14. Cans of corn are three for $.76. How much would 15 cans cost?

15. In a school with 320 students, 60% are boys. How many students are boys?

16. A man can pay for a car in 36 payments of $160 or pay $3,500 cash. How much can he save by paying in cash?

17. John earned 84 dollars and spent 21 dollars. What percent of his earnings did he spend?

18. Maria baked a pie and gave $\frac{1}{3}$ of it to Ellen and $\frac{2}{5}$ of it to Alfonso. What fraction of the pie did she have left?

19. Twelve people in a class passed the final exam. If this was 80% of the class, how many are in the class?

20. A man has a rectangular lot which is 36 feet by 20 feet. How many feet of fencing is needed to enclose it? If each four-foot section costs $35, how much will the fence cost?

Final Test - Whole Numbers

1.
$$\begin{array}{r} 342 \\ 53 \\ + 616 \end{array}$$

2.
$$\begin{array}{r} 746 \\ 716 \\ 823 \\ + 634 \end{array}$$

3. $7,362 + 775 + 72,516$

4. $7,013 + 2,615 + 776 + 29 =$

5. $7,001 + 696 + 18 + 732 =$

6.
$$\begin{array}{r} 743 \\ - 367 \end{array}$$

7.
$$\begin{array}{r} 5,282 \\ - 1,367 \end{array}$$

8. $7,052 - 2,637 =$

9. $6,000 - 3,678 =$

10. $7,001 - 678 =$

11.
$$\begin{array}{r} 76 \\ \times \quad 3 \end{array}$$

12.
$$\begin{array}{r} 7,653 \\ \times \quad 4 \end{array}$$

13.
$$\begin{array}{r} 53 \\ \times \quad 46 \end{array}$$

14.
$$\begin{array}{r} 627 \\ \times \quad 36 \end{array}$$

15.
$$\begin{array}{r} 673 \\ \times 346 \end{array}$$

16. $3 \overline{)425}$

17. $6 \overline{)1697}$

18. $30 \overline{)769}$

19. $42 \overline{)8992}$

20. $28 \overline{)1577}$

1.	
2.	
3.	
4.	
5.	
6.	
7.	
8.	
9.	
10.	
11.	
12.	
13.	
14.	
15.	
16.	
17.	
18.	
19.	
20.	
Score	

Final Test - Fractions

1. $\dfrac{3}{5}$

   $+\ \dfrac{1}{5}$

2. $\dfrac{5}{6}$

   $+\ \dfrac{3}{6}$

3. $\dfrac{2}{3}$

   $+\ \dfrac{1}{5}$

4. $3\dfrac{2}{3}$

   $+\ 4\dfrac{5}{9}$

5. $7\dfrac{3}{4}$

   $+\ 2\dfrac{3}{8}$

6. $\dfrac{5}{8}$

   $-\ \dfrac{1}{8}$

7. $7\dfrac{2}{5}$

   $-\ 2\dfrac{3}{5}$

8. 7

   $-\ 2\dfrac{3}{5}$

9. $6\dfrac{3}{4}$

   $-\ \dfrac{1}{2}$

10. $9\dfrac{1}{3}$

    $-\ 3\dfrac{2}{5}$

11. $\dfrac{2}{3} \times \dfrac{4}{7} =$

12. $\dfrac{12}{13} \times \dfrac{3}{24} =$

13. $\dfrac{3}{4} \times 36 =$

14. $\dfrac{7}{8} \times 2\dfrac{1}{7} =$

15. $2\dfrac{1}{3} \times 3\dfrac{1}{2}$

16. $\dfrac{3}{4} \div \dfrac{1}{2} =$

17. $3\dfrac{1}{2} \div \dfrac{1}{2} =$

18. $3\dfrac{2}{3} \div 1\dfrac{1}{2} =$

19. $3\dfrac{3}{4} \div 1\dfrac{1}{8} =$

20. $6 \div 2\dfrac{1}{3} =$

1.
2.
3.
4.
5.
6.
7.
8.
9.
10.
11.
12.
13.
14.
15.
16.
17.
18.
19.
20.
Score

Final Test - Decimals

1. $\begin{array}{r} 3.72 \\ 4.6 \\ +\ 3.963 \\ \hline \end{array}$

2. $.3 + 2.96 + 7.1 =$

3. $15.4 + 4 + 9.7 =$

4. $\begin{array}{r} 37.3 \\ -\ 16.7 \\ \hline \end{array}$

5. $\begin{array}{r} 7.1 \\ -\ 2.37 \\ \hline \end{array}$

6. $6 - 1.43 =$

7. $\begin{array}{r} 3.12 \\ \times\ \ \ \ 3 \\ \hline \end{array}$

8. $\begin{array}{r} 3.4 \\ \times\ \ 16 \\ \hline \end{array}$

9. $\begin{array}{r} .47 \\ \times\ 1.6 \\ \hline \end{array}$

10. $\begin{array}{r} .436 \\ \times\ 3.21 \\ \hline \end{array}$

11. $100 \times 2.36 =$

12. $1,000 \times 2.7 =$

13. $2\overline{)2.68}$

14. $5\overline{)7.3}$

15. $.5\overline{).325}$

16. $.003\overline{)1.2}$

17. $.15\overline{).0075}$

18. $8.7\overline{).1131}$

19. Change $\dfrac{7}{8}$ to a decimal.

20. Change $\dfrac{11}{25}$ to a decimal.

1.	
2.	
3.	
4.	
5.	
6.	
7.	
8.	
9.	
10.	
11.	
12.	
13.	
14.	
15.	
16.	
17.	
18.	
19.	
20.	
Score	

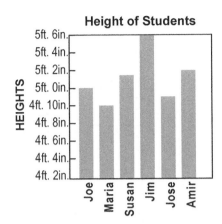

Height of Students

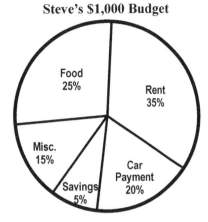

Steve's $1,000 Budget

1. Which student is 5' 0"?

2. Who are the two tallest students?

3. Who is the fourth tallest student?

4. What is the combined height of Maria and Joe?

5. How many inches taller is Jim than Maria?

6. What percent of the budget is spent on rent?

7. What percent is spent on food?

8. How many dollars are spent on rent? (Hint: Find 35% of $1,000.)

9. How many dollars are saved per month?

10. What percent is left after food and rent?

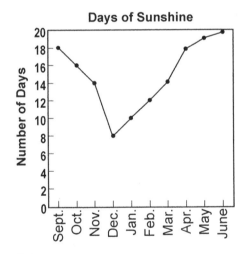

Days of Sunshine

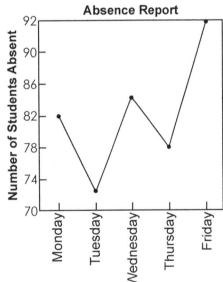

Absence Report

11. How many days of sunshine were there in December?

12. How many more sunny days were there in May than in March?

13. Which month had the third fewest sunny days?

14. What is the total number of sunny days in October and June?

15. Which month had the second most sunny days?

16. Which day has the most absences?

17. Which day had the most students present?

18. Which day had the fewest absences?

19. What was the total number of students absent on Monday and Thursday?

20. Which two days had the most absences?

1.

2.

3.

4.

5.

6.

7.

8.

9.

10.

11.

12.

13.

14.

15.

16.

17.

18.

19.

20.

Score

Final Test - Problem Solving

1. Mt. Baxter is 12,496 feet high and Mt. Henry is 13,998 feet high. How much higher is Mt. Henry than Mr. Baxter?

2. A class has 35 students. If $\frac{2}{5}$ of them are boys, how many girls are in the class?

3. A plane traveled 485 miles per hour for nine hours. What was the total distance traveled?

4. A rope nine feet long was cut into $1\frac{1}{2}$ - foot pieces. How many pieces were there?

5. Six pounds of corn cost $4.14. What is the price for two cans?

6. A shirt cost $15. If the sales tax was 8%, what was the total cost of the shirt?

7. Manuel scored a total of 344 points on four tests. What was his average score?

8. A race trace is $2\frac{1}{2}$ miles long. How far would you travel if you drove 12 laps around the track?

9. A tailor has nine yards of cloth. He cut two pieces which were $1\frac{2}{5}$ yards each. How much cloth was left?

10. Will is taking a test with 45 problems and he has finished $\frac{3}{5}$ of them. How many problems are left?

1.
2.
3.
4.
5.
6.
7.
8.
9.
10.
Score

11. A boy bought four sodas for $.79 each and five burgers for $1.19 each. What was the total cost?

12. Four people paid $60.80 for dinner and left a 20% tip. If they divided the cost evenly, how much did each pay?

13. Three tickets to a movie cost $9.75. How much will twelve tickets cost?

14. A man bought a car for $24,000. If he made a $4,000 down payment and paid the rest in equal payments of $500, how many payments would he make?

15. A rancher had 500 pounds of beef. He gave $\frac{2}{5}$ of it away and put $\frac{1}{2}$ of the remainder in his freezer. He sold the rest for $4.00 per pound. How much did he receive for the sale of the beef?

16. A team played 15 games and won 12 of them. What percent of the games did the team lose?

17. A school has 350 boys and 320 girls. If $\frac{3}{5}$ of the boys take French and $\frac{5}{8}$ of the girls take French, how many students are taking French?

18. Ferdie bought a CD for $12.00 and a DVD for $15.00. What is the total cost if sales tax is 8%?

19. Cans of soda are three for $1.68. How much would two sodas and a bag of chips priced at $1.79 cost altogether?

20. Dean took a test with 20 problems and got 19 of them correct. What percent did he get correct?

11.
12.
13.
14.
15.
16.
17.
18.
19.
20.
Score

Solutions

Page 4

Review Exercises
1. 605
2. 684
3. 1,130
4. 1,944
5. 30,018
6. 44,674

S1. April
S2. 10°
1. July
2. July
3. June
4. 7° - 8°
5. August
6. March; April
7. May; July
8. 13°
9. 3° - 4°
10. March; April; July

Page 5

Review Exercises
1. 1,520
2. 427
3. 4,494
4. 1,075
5. 940
6. 2,508

S1. Auberry; Winston
S2. 1,050
1. 1,200
2. 50
3. 350
4. 3,050 - 3,075
5. Approx. 325
6. 450
7. Approx. 1,375
8. 50
9. Sun City
10. 1,700

Page 6

Review Exercises
1. 15,630
2. 1,472
3. 9,541
4. 3,772
5. 205 r2
6. 47 r1

S1. 90
S2. 10
1. Tests 4, 6, and 7
2. 20
3. Approx. 91
4. 4
5. 80, 85
6. 90
7. 5
8. 15
9. 3
10. improved

Page 7

Review Exercises
1. 9,872
2. 4,278
3. 11,310
4. 201
5. 399 r3
6. 180 r5

S1. July; September
S2. 100
1. 700
2. 100
3. May
4. June
5. 100
6. July; September
7. 1,100
8. 200
9. 100
10. April; May

Page 8

Review Exercises
1. 15,272
2. 4,144
3. 613
4. 3 r6
5. 15 r 15
6. 201

S1. 23%
S2. 70%
1. 33%
2. $460
3. $200
4. $400
5. 68%
6. 32%
7. $24,000
8. 80%
9. car, clothing
10. other expenses

Page 9

Review Exercises
1. 38
2. 145
3. 11 r25
4. 9 r67
5. 117 r6
6. 205 r2

S1. 4/24 = 1/6
S2. 6/24 = 1/4
1. 5
2. 10
3. 8
4. 9/24 = 3/8
5. 16/24 = 2/3
6. 90
7. 6:00 A.M.
8. 2:30 P.M.
9. 40
10. 6/24 = 1/4

Page 10

Review Exercises
1. 1,074
2. 5,644
3. 89,790
4. 21 r1
5. 21
6. 21 r1

S1. 6,000
S2. 4,500
1. 1989
2. 13,500
3. 1989; 1991
4. 7,000
5. 11,000
6. $150,000
7. $600,000
8. 5,000
9. 19,500
10. 3,000

Page 11

Review Exercises
1. 17 r1
2. 241 r6
3. 23 r7
4. 71 r27
5. 349 r17
6. 225 r11

S1. 1990
S2. approx. 6-7 hours
1. 45
2. approx 5
3. 2,000
4. 1960
5. 1950; 1960
6. approx. 2
7. 8
8. approx. 12
9. approx. 5
10. approx 10

Solutions

Page 12

Review Exercises
1. Ken
2. 70
3. 40
4. 200
5. Sue
6. 30%
7. 25%
8. 55%
9. Food
10. 15%
11. 60
12. cat
13. fish
14. 30
15. dog
16. 400
17. August
18. 400
19. 500
20. July; August

Page 13

Review Exercises
1. Grade 4
2. Approx. 1,500 cans
3. Approx. 9,500 cans
4. Approx. 21,500 cans
5. Grade 2
6. 20 students
7. 11 more C's
8. 100 students
9. 13 more C's and B's
10. B's
11. 25 inches
12. Approx. 9 inches
13. January
14. 60 inches
15. December
16. 500 boxes
17. troop 15
18. 200 boxes
19. 900 boxes
20. 750 boxes

Page 14

Review Exercises
S1. 29,710
S2. 3,655
1. 547
2. 573
3. 10,103
4. 7,243,008
5. 5,665
6. 1,196
7. 3,418
8. 1,845
9. 616
10. 90,558

Page 15

Review Exercises
S1. 29,025
S2. 220 r9
1. 228
2. 30,612
3. 22,572
4. 232,858
5. 141 r2
6. 282 r5
7. 25 r19
8. 94 r2
9. 214 r4
10. 56 r9

Page 16

Review Exercises
1. 1,236
2. 7,163
3. 2,127
4. 391
5. 58 r1
6. 195 r8

S1. 138
S2. 1,580
1. 391 increase
2. $139
3. 14,532 feet
4. 55 mph
5. 682 miles

Page 17

Review Exercises
1. 9,152
2. 207,936
3. 424
4. 13 r15
5. 118 r43
6. 142 r23

S1. 173 pieces
S2. 5 buses
1. $27
2. 21,500 gallons
3. 125 pounds
4. 1,650 cars
5. 450 mph

Page 18

Review Exercises
1. 1 r 14
2. 17 r22
3. 237 r22
4. 2 r1
5. 32 r5
6. 315 r1

S1. $705
S2. $312
1. 416 votes
2. 866 feet
3. $27
4. 197 containers
5. 840 students

Page 19

Review Exercises
1. 632
2. 5,705
3. 305
4. 124
5. 2,982
6. 6,754

S1. 1,032
S2. $640
1. 10,940
2. 180 cows
3. 1,219 students
4. 24 students
5. 89

Solutions

Page 20

Review Exercises
1. 55 mph
2. 4,368
3. 750 students
4. 1,259
5. $1,061
6. 6,437

S1. 90
S2. 1,985 pounds
1. $20,280
2. 117 hours
3. 356 seats
4. 186,650 gallons
5. $42

Page 21

Review Exercises
1. 14,766
2. 86,892 books
3. 23 r15
4. 5,460 miles
5. 164 r8
6. $6,695

S1. $385
S2. 8 boxes, 8
1. 72 seats
2. 4 pieces
3. $97
4. 24 classes
5. 2,064 hours

Page 22

Review Exercises
1. 18 boxes
2. 6,199
3. 116 feet
4. 7,280
5. 376, 94
6. 339 r2

S1. $8
S2. 17 miles
1. 470 feet
2. 720 bushels
3. $1,200
4. 4 buses
5. 87

Page 23

Review Exercises
1. 21
2. 50 students
3. 1,261
4. $13
5. 42,350
6. 95

S1. $148
S2. $250
1. $2,720
2. $15,850
3. 1,485 miles
4. 51,925 gallons
5. $89,500

Page 24

Review Exercises
1. 80
2. 11,421
3. 384 students
4. 201
5. 1,920
6. 8,722

S1. $39
S2. 80 feet; $560
1. 480 miles
2. 17 hrs. 10 min.
3. $600
4. $900
5. $163

Page 25

Review Exercises
1. 9,732
2. 550 mph
3. 85
4. 2,180
5. $36
6. $1,618
7. 18 classes
8. 30 payments
9. 272 feet; $850
10. $2,520

Page 26

Review Exercises
S1. 1 3/7
S2. 10 1/4
1. 7/18
2. 13/18
3. 4 2/5
4. 8 5/8
5. 11 1/4
6. 4 3/4
7. 8 7/15
8. 3 1/2
9. 11/16
10. 8 7/12

Page 27

Review Exercises
S1. 3
S2. 3 2/3
1. 3/26
2. 27
3. 1 7/8
4. 8 1/6
5. 1 1/2
6. 7
7. 2 4/9
8. 3 1/3
9. 2 4/7
10. 1 1/3

Page 28

Review Exercises
1. 5/6
2. 3 13/15
3. 5 1/6
4. 2 1/6
5. 5 2/15
6. 5 4/5

S1. 5 3/4 cups
S2. $40
1. 8 1/2 pounds
2. 11 pieces
3. 30 miles
4. 29 1/4 minutes
5. 34 feet

Page 29

Review Exercises
1. 2/5
2. 5/6
3. 3
4. 1 1/3
5. 9
6. 1 3/4

S1. 15 pounds
S2. 1 1/2 pounds
1. 14 1/4 hours
2. 125 miles
3. 16 tires
4. 27 pounds
5. 4 3/4 hours

Page 30

Review Exercises
1. 1 1/10
2. 5/8
3. 6 1/6
4. 6 2/15
5. 5 1/10
6. 3 7/10

S1. $78
S2. 12 packages
1. 1 1/4 hours
2. 63 1/4 inches
3. 35 1/6 pages
4. 13 1/3 pounds
5. 90 miles

Page 31

Review Exercises
1. 4/11
2. 11
3. 2
4. 1 1/2
5. 1 1/3
6. 2

S1. 33 feet
S2. 30 pounds
1. 5 1/5 feet
2. 1/2 pound
3. 3 5/6 miles
4. 27 1/2 feet
5. $2,100

Solutions

Page 32

Review Exercises
1. 1 3/4 inches
2. 1 1/12
3. 21 pages
4. 5/8
5. 6 1/4 pounds
6. 5 1/4

S1. 4 yards
S2. $14
1. 1 1/4 gallons
2. 18 girls
3. $147
4. 24 bracelets
5. 2,000 acres

Page 33

Review Exercises
1. 14
2. 11 patties
3. 3 1/3
4. 10 yards
5. 3
6. 23 1/12 minutes

S1. 60 miles
S2. $48
1. 17 bushels
2. 1 1/2 pounds
3. $120,000
4. $2,600
5. 112 pages

Page 34

Review Exercises
1. 1 7/8
2. $48
3. 1,299
4. 520 miles
5. 21 r9
6. 160 11/12 pounds

S1. 1/6
S2. 406 pages
1. 5 1/2 miles
2. 40, $80
3. 1 1/2 pounds
4. 27 3/4 hours
5. $20

Page 35

Review Exercises
1. 6 1/4
2. 8/15
3. 1 11/16
4. 8 3/4
5. 2 1/4
6. 1 7/8

S1. $19
S2. 9 3/4 dollars
1. $36
2. $51
3. 6 pounds
4. $8
5. $840

Page 36

Review Exercises
1. 15
2. 28 r16
3. 7 1/2
4. 11,102
5. 7/30
6. 5,409

S1. 23 students
S2. 352 lots,
    $1,760,000
1. 510 students
2. 200 acres
3. $800
4. $240
5. $60,000

Page 37

Review Exercises
1. $48
2. 116 3/4 pounds
3. 8 pieces
4. 10 5/12 hours
5. $147
6. $47
7. 20 packages, $100
8. 150; 100
9. 1/6 pie
10. $725

Page 38

Review Exercises
S1. 18.38
S2. 3.687
1. 23.7114
2. 6.17
3. 16.3
4. 8.24
5. 2.08
6. 19.56
7. 2.2
8. 4.487
9. 10.396
10. 25.7

Page 39

Review Exercises
S1. 7.452
S2. .65
1. 12.78
2. 54.4
3. 4.928
4. 147.888
5. 1.34
6. 1.46
7. 400
8. .75
9. .05
10. .013

Page 40

Review Exercises
1. 11.243
2. 20.95
3. 4.478
4. 2.636
5. 6.42
6. 18.56

S1. $63.60
S2. 500 mph
1. $.34
2. $518.85
3. 194 feet
4. $1.14
5. $12.60

Page 41

Review Exercises
1. .216
2. .0786
3. .000006
4. 1.84
5. .026
6. .007

S1. 360 miles
S2. 7 pounds
1. 278.35 pounds
2. $.89
3. 11.2 gallons
4. 20 cans
5. $3.36

Page 42

Review Exercises
1. 6.7
2. 6.5
3. 170
4. 5.4
5. 9.2
6. 3.2

S1. $15.84
S2. $47.70
1. $2.25
2. $1931.24
3. $75.48
4. $250.50
5. $9.84

Page 43

Review Exercises
1. 13.24
2. 3.287
3. 2.562
4. $22
5. $3.15
6. 250

S1. $620
S2. 41.6 inches
1. $723.59
2. $35.15
3. 13.53 inches
4. $3.20
5. $3.40

Solutions

Page 44

Review Exercises
1. .6
2. $30
3. 14/15
4. 2 1/2 feet
5. 1 9/14
6. 30 inches

S1. $13.40
S2. $8.46
1. $24.80
2. $625.75
3. $7.14
4. $5.10
5. $7.05

Page 45

Review Exercises
1. 1825 miles
2. 1 1/7
3. 408 engines
4. 9 1/3
5. 2 1/2 feet
6. 2

S1. $12.84
S2. $37.79
1. $2.62
2. 12.1
3. $18
4. $8.95
5. $468

Page 46

Review Exercises
1. 53 mph
2. 2,656
3. 3,504
4. 7,982
5. 15 miles
6. 1,193

S1. 25.02
S2. 19.32
1. 5.31
2. 16.884
3. $504
4. $14.95
5. $1.25

Page 47

Review Exercises
1. 11.22
2. 4.835
3. 21.288
4. .26
5. .05
6. 5,000

S1. 7.9 miles
S2. $5,600,000
1. $45
2. $4.70
3. $147
4. $170
5. $44.79

Page 48

Review Exercises
1. 5 1/2
2. 1 3/8
3. 3 7/12
4. 9 3/40
5. 4 3/4
6. 4 3/4

S1. $38.79
S2. $960
1. 38 students
2. $79.92
3. 480 acres
4. 161.5 miles
5. $1.28

Page 49

1. 448 miles
2. $1.19
3. 7.8 pounds
4. $388.94
5. 9 pounds
6. $44.10
7. $19.34
8. $13.80
9. $10,500,000
10. $7.65

Page 50

S1. .2; 1/50
S2. .09; 9/100
1. .16; 4/25
2. .06; 3/50
3. .75; 3/4
4. .4; 2/5
5. .01; 1/100
6. .45; 9/20
7. .12; 3/25
8. .05; 1/20
9. .5; 1/2
10. .13; 13/100

Page 51

S1. 20%
S2. 20
1. 42
2. 21
3. 25%
4. 25
5. 20
6. 60
7. 75%
8. 16
9. 60
10. 20%

Page 52

Review Exercises
1. 433.8
2. 2.382
3. 7.2
4. 17.5
5. 15
6. 5

S1. 32 problems
S2. 470 students
1. $1,800
2. 9 stamps
3. $30,000
4. $4.20; $64.20
5. $12,000

Page 53

Review Exercises
1. 4.8
2. 48
3. 60%
4. 60%
5. 75%
6. 90%

S1. 1,500 acres
S2. $297.50
1. 180 girls
2. $17.28
3. 36 pages
4. $6.00
5. 32 games

Page 54

Review Exercises
1. 40%
2. 75%
3. 25%
4. 90%
5. 24
6. 9.6

S1. 80%
S2. 40%
1. 75%
2. 84%
3. 75%
4. 25%
5. 60%

Page 55

Review Exercises
1. 5.4
2. 80%
3. 15
4. 26
5. 30%
6. 24

S1. 75%
S2. 40%
1. 80%
2. 30%
3. 75%
4. 40%
5. 60%

Solutions

Page 56

Review Exercises
1. 15
2. 160
3. 12
4. 120
5. 20%
6. 90%

S1. 25 games
S2. 120 stamps;
    96 stamps
1. $30
2. 35
3. 1,000 bushels
4. 12 shots; 3 shots
5. 30 problems

Page 57

Review Exercises
1. 20
2. 40%
3. 225
4. 25
5. 80%
6. 22.5

S1. 50 questions
S2. 125 people
1. 20 marbles;
   12 marbles
2. $25,000
3. 15
4. 50 students
5. 4 games

Page 58

1. 55
2. 30
3. 20%
4. 95%
5. 8 people
6. 75%
7. 35 students
8. 75%
9. 2,400 acres
10. 80%

Page 59

1. 60%
2. 40
3. 25%
4. 67.5
5. 25 games
6. 180 students
7. 75%
8. 56 gold fish
9. 75%
10. $34.56

Page 60

1. 814
2. 2,178
3. 23,965
4. 9,314
5. 7,693
6. 279
7. 3,628
8. 1,414
9. 1,714
10. 5,284
11. 292
12. 28,544
13. 1,665
14. 15,732
15. 158,178
16. 131 r2
17. 344
18. 14 r8
19. 253 r24
20. 69 r1

Page 61

1. 5/7
2. 1 1/4
3. 14/15
4. 5 7/8
5. 14 3/10
6. 3/4
7. 3 1/2
8. 2 6/7
9. 5 1/10
10. 4 11/12
11. 6/35
12. 3/44
13. 20
14. 2
15. 8 3/4
16. 2 1/2
17. 4 2/3
18. 1 1/3
19. 3 2/3
20. 4 1/2

Page 62

1. 11.913
2. 14.52
3. 30.7
4. 18.6
5. 2.61
6. 70.32
7. 7.92
8. 189.8
9. 1.512
10. .53298
11. 36.5
12. 3,600
13. 2.32
14. 1.34
15. .31
16. 300
17. .03
18. 13
19. .6
20. .35

Page 63

1. Grand Falls
2. 525 - 550 feet
3. 100 feet
4. Snake Falls
5. Morton Falls
6. 13%
7. 20%
8. $600
9. $300
10. 41%
11. 71° - 72°
12. 30°
13. June
14. August
15. 10°
16. 60,000
17. 40,000
18. 200,000
19. 180,000 pounds
20. salmon, cod,
    and snapper

Solutions

Page 64

1. $3,514
2. 603
3. 50 problems
4. 12 bags
5. 3 3/4 dollars
6. 208 miles
7. $20.25
8. 103 pounds
9. $342.75
10. $2.81

Page 65

11. 28 correct
12. $30
13. $21.65
14. $3.80
15. 192 boys
16. $2,260
17. 25%
18. 4/15 pie
19. 15 students
20. 112 feet; $980

Page 66

1. 1,011
2. 2,919
3. 80,653
4. 10,433
5. 8,447
6. 376
7. 3,915
8. 4,415
9. 2,322
10. 6,323
11. 288
12. 30,612
13. 2,438
14. 22,572
15. 232,858
16. 141 r2
17. 282 r5
18. 25 r19
19. 214 r4
20. 56 r9

Page 67

1. 4/5
2. 1 1/3
3. 13/15
4. 8 2/9
5. 10 1/8
6. 1/2
7. 4 4/5
8. 4 2/5
9. 6 1/4
10. 5 14/15
11. 8/21
12. 3/26
13. 27
14. 1 7/8
15. 8 1/6
16. 1 1/2
17. 7
18. 2 4/9
19. 3 1/3
20. 2 4/7

Page 68

1. 12.283
2. 10.36
3. 29.1
4. 20.6
5. 4.73
6. 4.57
7. 9.36
8. 54.4
9. .752
10. 1.39956
11. 236
12. 2,700
13. 1.34
14. 1.46
15. .65
16. 400
17. .05
18. .013
19. .875
20. .44

Page 69

1. Joe
2. Amir, Jim
3. Joe
4. 9' 10"
5. 8 inches
6. 35%
7. 25%
8. $350
9. $50
10. 40%
11. 8
12. 4-5
13. February
14. 36
15. May
16. Friday
17. Tuesday
18. Tuesday
19. 160-170 Appr.
20. Wednesday, Friday

Page 70

1. 1,502 feet
2. 21 girls
3. 4,365 miles
4. 6 pieces
5. $1.38
6. $16.20
7. 86
8. 30 miles
9. 6 1/5 yards
10. 18 problems

Page 71

11. $9.11
12. $18.24
13. $39
14. 40 payments
15. $600
16. 20%
17. 410 students
18. $29.16
19. $2.91
20. 95%

Made in the USA
Monee, IL
21 January 2021